life science data mining

SCIENCE, ENGINEERING, AND BIOLOGY INFORMATICS

Series Editor: Jason T. L. Wang
               (*New Jersey Institute of Technology, USA*)

---

life science data mining

editors

Stephen Wong
Harvard Medical School, USA

Chung-Sheng Li
IBM Thomas J Watson Research Center

 **World Scientific**

NEW JERSEY · LONDON · SINGAPORE · BEIJING · SHANGHAI · HONG KONG · TAIPEI · CHENNAI

Published by

World Scientific Publishing Co. Pte. Ltd.

5 Toh Tuck Link, Singapore 596224

USA office: 27 Warren Street, Suite 401-402, Hackensack, NJ 07601

UK office: 57 Shelton Street, Covent Garden, London WC2H 9HE

British Library Cataloguing-in-Publication Data
A catalogue record for this book is available from the British Library.

Science, Engineering, and Biology Informatics — Vol. 2
LIFE SCIENCE DATA MINING

ISBN-13 978-981-270-064-3
ISBN-10 981-270-064-1
ISBN-13 978-981-270-065-0 (pbk)
ISBN-10 981-270-065-X (pbk)

Printed in Singapore

PREFACE

Data mining is the process of using computational algorithms and tools to automatically discover useful information in large data archives. Data mining techniques are deployed to score large databases in order to find novel and useful patterns that might otherwise remain unknown. They also can be used to predict the outcome of a future observation or to assess the potential risk in a disease situation. Recent advances in data generation devices, data acquisition, and storage technology in the life sciences have enabled biomedical research and healthcare organizations to accumulate vast amounts of heterogeneous data that is key to important new discoveries or therapeutic interventions. Extracting useful information has proven extremely challenging however. Traditional data analysis and mining tools and techniques often cannot be used because of the massive size of a data set and the non-traditional nature of the biomedical data, compared to those encountered in financial and commercial sectors. In many situations, the questions that need to be answered cannot be addressed using existing data analysis and mining techniques, and thus, new algorithms and methods need to be developed.

Life science is an important application domain that requires new techniques of data analysis and mining. This is one of the first technical books focusing on the data analysis and mining techniques in life science applications. In this introductory chapter, we present the key topics to be covered in this book. In Chapter 1 "Taxonomy of early detection for environmental and public health applications," Chung-Sheng Li of IBM Research provides a survey of early warning systems and detection approaches in terms of problem domains and data sources. The chapter introduces current syndromic surveillance prototypes or deployments and defines the problem domain for three classes: individual and public health level, cellular level, and molecular level. For data sources, they were also categorized into three parts including clinically related data, non-traditional data, and auxiliary data. Furthermore, data sources can be characterized by three dimensions (structured, semi-structured, and non-structured).

In Chapter 2 "Time-lapse cell cycle quantitative data analysis using Gaussian mixture models," Xiaobo Zhou and colleagues at Harvard Medical School describe an interesting and important emerging technology area of high throughput biological imaging. The authors address the unresolved problem of identifying the cell cycle process under different conditions of perturbation. In the study, the time-lapse fluorescence microscopy imaging images are analyzed to detect and measure the duration of various cell phases, e.g., inter phase, prophase, metaphase, and anaphase, quantitatively.

Chapter 3 "Diversity and accuracy of data mining ensemble" by Wenjia Wang of the University of East Anglia discusses an important issue: classifier fusion or an ensemble of classifiers. This paper first describes why diversity is essential and how the diversity can be measured, then it analyses the relationships between the accuracy of an ensemble and the diversity among its member classifiers. An example is given to show that the mixed ensembles are able to improve the performance.

Various clustering algorithms have been applied to gene expression data analysis. Chapter 4 "Integrated clustering for microarray data" by Gabriela Moise and Jorg Sander from the University of Alberta argue that the integration strategy is a "majority voting" approach based on the assumption that objects belong to a "natural" cluster are likely to be co-located in the same cluster by different clustering algorithms. The chapter also provides an excellent survey of clustering and integrated clustering approaches in microarray analysis.

EEG has a variety of applications in basic and clinic neuroscience. Chapter 5: "Complexity and Synchronization of EEG with Parametric Modeling" by Xiaoli Li presents a nice work on parametric modeling of the complexity and synchronization of EEG signals to assist the diagnosis of epilepsy or the analysis of EEG dynamics.

In Chapter 6: "Bayesian Fusion of Syndromic Surveillance with Sensor Data for Disease Outbreak Classification," by Jeffrey Lin, Howard Burkom, *et al.* from The Johns Hopkins University Applied Physics Laboratory and Walter Reed Army Institute for Research describe a novel Bayesian approach to fuse sensor data with syndromic surveillance data presented for timely detection and classification of disease outbreaks. In addition, the authors select a natural disease, asthma, which is highly dependent on environmental factors to validate the approach.

Continuing the theme of syndromic surveillance, in Chapter 7: "An Evaluation of Over-the-Counter Medication Sales for Syndromic Surveillance," Murray Campbell, Chung Sheng Li, *et al.* from IBM Watson Research Center describe a number of approaches to evaluate the utility of data sources in a syndromic surveillance context and show that there may be some values in using sales of over-the-counter medications for syndromic surveillance.

In Chapter 8: "Collaborative Health Sentinel" JH Kaufman, G Decad, *et al.* from IBM Research Divisions in California, New York, and Israel provide a clear survey and highlight the approach of significant trends, issues and further directions for global systems for health management. They addressed various issues for systems covering general environment and public health, as well as global view.

In Chapter 9: "Data Mining for Drug Abuse Research and Treatment Evaluation: Data Systems Needs and Challenges" Mary Lynn Brecht of UCLA argues that drug abuse research and evaluation can benefit from data mining strategies to generate models of complex dynamic phenomena from heterogeneous data sources. She presents a user's perspective and several challenges on selected topics relating to the development of an online processing framework for data retrieval and mining to meet the needs in this field.

The increasing amount and complexity of data used in predictive toxicology call for new and flexible approaches based on hybrid intelligent methods to mine the data. To fill this needs, in Chapter 10 "Knowledge Representation for Versatile Hybrid Intelligent Processing Applied in Predictive Toxicology", Daneil Neagu from Bradford University, England addresses the issue of devising a mark-up language, as an application of XML (Extensible Markup Language), for representing knowledge modeling in predictive toxicology - PToXML and the markup language HISML for integrated data structures of Hybrid Intelligent Systems.

Ensemble classification is an active field of research in pattern recognition. In Chapter 11: "Ensemble Classification System

Implementation for Biomedical Microarray Data," Shun Bian and Wenjia Wang of the University of East Anglia, UK present a framework of developing a flexible software platform for building an ensemble based on the diversity measures. An ensemble classification system (ECS) has been implemented for mining biomedical data as well as general data.

Time-lapse fluorescence microscopy imaging provides an important high throughput method to study the dynamic cell cycle process under different conditions of perturbation. The bottleneck, however, lies in the analysis and modeling of large amounts of image data generated. Chapter 12: "An Automated Method for Cell Phase Identification in High Throughput Time-Lapse Screens" by Xiaowei Chen and colleagues from Harvard Medical School describe the application of statistical and machine learning techniques to the problem of tracking and identifying the phase of individual cells in populations as a function of time using high throughput imaging techniques.

Modeling gene regulatory networks has been an active area of research in computational biology and systems biology. An important step in constructing these networks involves finding genes that have the strongest influence on the target gene. In Chapter 13, "Inference of Transcriptional Regulatory Networks based on Cancer Microarray Data," Xiaobo Zhou and Stephen Wong from Harvard Center for Neurodegeneration and Repair address this problem. They start with certain existing subnetworks and methods of transcriptional regulatory network construction and then present their new approach.

In Chapter 14: "Data Mining in Biomedicine," Lucila Ohno-Machado and Staal Vinterbo of Brigham and Women's Hospital provide an overview of data mining techniques in biomedicine. They refer to data mining as any data processing algorithm that aims to determine patterns or regularities in the data. The patterns may be used for diagnostic or prognostic purposes and the models that result from pattern recognition algorithms will be refereed to as *predictive models*, regardless of whether they are used to classify. The chapter provides readers an excellent introduction about data mining and its applications to the readers.

Association rules mining is a popular technique for the analysis of gene expression profiles like microarray data. In Chapter 15: "Mining

Multilevel Association Rules from Gene Ontology and Microarray Data," VS Tseng and SC Yang from National Cheng Kung University, Taiwan, aim at combining microarray data and existing biological network to produce multilevel association rules. They propose a new algorithm for mining gene expression transactions based on existing algorithm ML_T1LA in the context of Gene Ontology and a filter version CMAGO.

Optical biosensors are now utilized in a wide range of applications, from biological-warfare-agent detection to improving clinical diagnosis. In Chapter 16, "A Proposed Sensor-Configuration and Sensitivity Analysis of Parameters with Applications to Biosensors," HJ Halim from Liverpool JM University, England, introduces a configuration of sensor system and analytical model equations to mitigate the effects of internal and external parameter fluctuations.

The subject of data mining in life science, while relatively young compared to data mining in other application fields, such as finance and marketing, or to statistics or machine learning, is already too large to cover in a single book volume. We hope that this edition would provide the readers some of the specific challenges that motivate the development of new data mining techniques and tools in life sciences and serve as an introductory material to the researchers and practitioners interested in this exciting field of application.

Stephen TC Wong and Chung-Sheng Li

CONTENTS

CHAPTER 1

SURVEY OF EARLY WARNING SYSTEMS FOR ENVIRONMENTAL AND PUBLIC HEALTH APPLICATIONS

Chung-Sheng Li

IBM T. J. Watson Research Center, PO Box 704, Yorktown Heights, NY 10598

1. Introduction

Advances in sensors and actuators technologies and real-time analytics recently have accelerated the adoption of real-time or near real-time monitoring and alert generation for environmental and public health related activities.

Monitoring of environmental related activities include the use of remote sensing (satellite imaging and aerial photo) for tracking the impacts due to global climate change (e.g. detection of thinning polar ice as evidence of global warming, deforestation in the Amazon region), tracking the impact of natural and man-made disasters (e.g. earthquakes, tsunami, volcano eruptions, floods, and hurricane), forest fire, and pollution (including pollution of air, land, and water).

Early warnings for human disease or disorder at the individual level often rely on various symptoms or biomarkers. For example, retina scan has been developed as a tool for early diagnosis of diabetic retinopathy [Denninghoff 2000]. CA 125, a protein found in blood, has been used as the biomarker for screening early stages of ovarian cancer. The level of Prostate Specific Antigen (PSA) has been used for early detection of prostate cancer (usually in conjunction with a digital rectal exam). National Cancer Institute of NIH has sponsored the forming of the Early Detection Research Network (EDRN http://www3.cancer.gov/prevention/cbrg/edrn/) program since 1999. This comprehensive effort includes the development and validation of various biomarkers for detecting colon, liver, kidney, prostate, bladder, breast, ovary, lung, and pancreatic cancer.

Health and infectious disease surveillance for public health purposes have existed for many decades. Currently, World Health Organization (WHO) has set up a Global Outbreak Alert & Response Network (GOARN) to provide surveillance and response for communicable diseases around the world (http://www.who.int/csr/outbreaknetwork/en/). Examples of the diseases tracked by GOARN include anthrax, influenza, Ebola hemorrhagic fever, plague, smallpox, and SARS. The alert and response operations provided include epidemic intelligence, event verification, information management and dissemination, real time alert, coordinated rapid outbreak response, and outbreak response logistics. For the United States, the National Center for Infectious Diseases (NCID) at CDC has set up 30+ surveillance networks (see Section 2) to track individual infectious diseases or infectious disease groups.

Monitoring disease outbreaks for public health purposes based on environmental epidemiology has been demonstrated for a number of *vector-born infectious diseases* such as Hantavirus Pulmonary Syndrome (HPS), malaria, Lyme disease, West Nile virus (WNV), and Dengue fever [Glass 2001a, Glass 2001b, Glass 2002, Klein 2001a, Klein 2001b]. In these cases, environmental effects such as the change of ground moisture, ground temperature, and vegetation due to El Niño and other climate changes facilitate the population change of disease hosts (such as mosquitoes and rodents). These changes can then cause increased risk of diseases in the human population.

Recently, a new trend to track subtle human behavior changes due to disease outbreak has emerged to provide advanced warnings before significant casualties registered from clinical sources. This approach is also referred to in the literature as *syndromic surveillance* (or early detection of disease outbreaks, pre-diagnosis surveillance, non-traditional surveillance, enhanced surveillance, non-traditional surveillance, and disease early warning systems). This approach has received substantial interests and enthusiasm during the past few years, especially after Sept.

11, 2001 [Buehler 2003, Duchin 2003, Goodwin2004, Mostashari 2003, Pavlin 2003, Sosin 2003, and Wagner 2001].

In this chapter, we surveyed the most widely adopted disease surveillance techniques, early warning systems, and detection approaches in terms of problem domain and data sources. This survey intends to facilitate the clarification and discovery of interdependency among problem domains, data sources, and detection methods and thus facilitate sharing of data sources and detection methods across multiple problem domains, when appropriate.

2. Disease Surveillance

Health and infectious disease surveillance for public health purposes have existed for many decades. The following is a subset of the 30+ national surveillance systems coordinated by the National Center of Infectious Disease (NCID) at CDC (http://www.cdc.gov/ncidod/osr/site/surv_resources/surv_sys.htm).

Mortality Reporting System: from 122 cities & metropolitan areas, compiled by the CDC epidemiology program office;

Active Bacterial Core Surveillance (ABCs): at ten emerging infectious program sites such as California, Colorado, Connecticut, Georgia, etc.;

BaCon Study: on assessing the frequency of blood component bacterial contamination associated with transfusion;

Border Infectious Disease Surveillance Project (BIDS): focusing on hepatitis and febrile-rash illness along the US-Mexico border;

Dialysis Surveillance Network (DSN): focus on tracking vascular access infections and other bacterial infections in hemodialysis patients;

Electronic Foodborne Outbreak Investigation and Reporting System (EFORS): used by 50 states to report data about Foodborne Outbreaks on a daily basis;

EMERGEncy ID NET: focus on emerging infectious disease based on 11 university affiliated urban hospital emergency departments;

Foodborne Diseases Active Surveillance Network (FoodNet): consists of active surveillance for foodborne diseases and related epidemiologic studies. This is a collaboration among CDC, the ten emerging infectious program sites (EIPs), the USDA, and FDA;

Global Emerging Infectious Sentinel Network (GeoSentinel): consists of travel/tropical medicine clinics around the world that monitor geographic and temporal trends in mobility;

National Notifiable Diseases Surveillance System (NNDSS): The Epidemiology Program Office (EPO) at CDC collect, compile, and publish reports of disease considered as notifiable at the national level;

National West Nile Virus Surveillance System: Since 2000, 48 states and 4 cities started to collect data about wild birds, sentinel chicken flocks, human cases, veterinary cases, and mosquito surveillance;

Public Health Laboratory Information System (PHLIS): Collects data on cases/isolates of specific notifiable diseases from every state within the United States.

United States Influenza Sentinel Physicians Surveillance Network: 250+ physicians around the country report each week the number of patients seen and the total number with flu-like symptoms;

Waterborne-Disease Outbreak Surveillance System: includes data regarding outbreaks associated with drinking water and recreational water.

The US DoD Global Emerging Infectious Surveillance System (DoD-GEIS): has its own set of surveillance network within the States as well as internationally (http://www.geis.fhp.osd.mil/).

A number of syndromic surveillance prototypes or deployments have been developed recently:

BioSense: a national initiative at US (coordinated by CDC) to establish near real-time electronic transmission of data to local, state, and federal public health agencies from national, regional, and local health data by accessing and analyzing diagnostic and prediagnostic health data.

RODS (Real Time Outbreak and Disease Surveillance http://rods.health.pitt.edu/): a joint effort between University of Pittsburgh and CMU. This effort includes an open source version of RODS software (since 2003) and National Retail Data Monitor (NRDM) which monitors the sales of over-the-counter healthcare products in 20,000 participating pharmacies in the states.

The New York City Department of Health and Mental Hygiene [Greenko 2003] has established a syndromic surveillance system since 2001 that monitors emergency department visits to provide early detection of disease outbreaks. The key signs and symptoms analyzed include respiratory problems, fever, diarrhea, and vomiting.

Concerns do exist for the effectiveness and general applicability of syndromic surveillance [Stoto 2004]. In particular, the size and timing of an outbreak for which syndromic surveillance gives an early advantage may be limited. In a case involving hundreds or thousands of simultaneous infections, no special detection methods would be needed. Conversely, even the best syndromic surveillance system might not work in a case involving only a few individuals such as the anthrax episodes of 2001 in the United States. Consequently, the size of an outbreak in which syndromic surveillance can provide added value (on top and beyond the traditional infectious disease surveillance) is probably on the order of tens to several hundred people.

3. Reference Architecture for Model Extraction

Figure 1 illustrates the general reference architecture of an end-to-end surveillance and response system. The surveillance module is responsible for collecting multi-modal data from available data sources, and conduct preliminary analysis based on one or more data sources. The surveillance module works closely with the simulation and modeling module to extract the model from these data sources. The primary responsibility for the simulation and modeling module is to determine whether an anomaly has already happened or is likely to happen in the future. Furthermore, it is also responsible for interacting with the decision maker to provide decision support on multiple what-if scenarios. The decision maker can then use the command and control module to launch appropriate actions (such as deploy antibiotics, applying ring vaccination, and quarantine) based on the decisions.

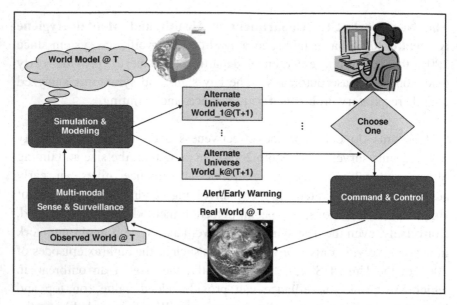

Figure 1: Reference architecture of an end-to-end surveillance & response system.

Surveillance and activity monitoring in environmental and public health domain often involves (1) collecting data from clinical data (patient records, billing information, lab results, etc.) as well as both hardware and/or software sensors which generate multi-modal multi-scale spatio-temporal data, (2) cleaning and preprocessing the data to remove gaps and inconsistencies, if any, and (3) applying anomaly detection methodologies to generate predictions and early warnings. Predictions are often based on the environmental and public health threat models.

As an example, we might subconsciously or consciously adjust our diet when we feel ill (such as drinking more water, juice, and have more rest).   If the symptoms become more severe, we might seek over-the-counter (OTC) medicine, and miss classes/work. In many cases, we might go to work late or leave for home early. We might also experience subtle change of our behavior at work. When the symptoms continue, we might seek help from physicians. We usually have to make appointments with the physicians (e.g. through phone or through web) first. Physician visits often result in prescription medicines and thus visits to the pharmacies.

This behavior model suggests that we could potentially observe the progression of a disease outbreak within a population at multiple touch points by this model – such as sales of drinking water, sewage generation, OTC drug sales, absentee information from school and work, phone records to physician office, and 911 calls.   Some of these data may be collected before visits to the physician or hospital have actually happened, and thus enable earlier detection. Data can also be collected during and after visits to physicians have been made. These data could include diagnosis made by the physician (including chief complaints), lab tests, and drug prescriptions. Similar models can also be established for pollution, non-infectious diseases, and other natural disasters.

We assume that the real world phenomenon, whether it is a disease outbreak or an environmental event, will induce noticeable changes in human, animal, or other environmental behaviors.   The behavioral change can be potentially captured from various hardware and/or software sensors as well as traditional business transactions.   The challenge for early warning is to fuse those appropriate multi-modal and multi-scale data sources to extract the behavior model of human or environment and compared to the behavior under normal conditions (if known) and declare the existence of anomaly when the behavior is believed to be abnormal.

Figure 2 shows that the detection and early warning approaches in environmental and public health application can be characterized along three dimensions: problem domain, data sources, and detection methodologies, each of which will be further elaborated in the following three sections.   These three dimensions are also the key aspects for materializing the reference architecture shown in Fig. 1 for any specific application. These three dimensions are often strongly correlated – as a given problem domain (such as detecting influenza outbreak) will dictates the available data sources and the most suitable detection methodologies.

Figure 3 shows the reference architecture of a Health Activity Monitoring system that materializes the multi-modal sense/surveillance and modeling modules in Fig. 1. Data from various sources (both traditional and non-traditional data) are streaming through the messaging

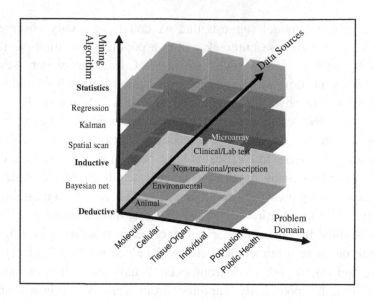

Figure 2: Taxonomy of data mining issues.

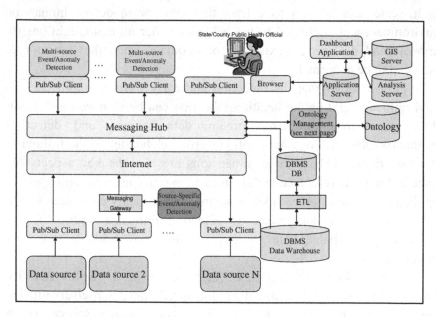

Figure 3: Reference Architecture for Health Activity Monitoring.

hub to anomaly detectors. A messaging hub is responsible for delivering the messages and events from the publishers to the intended subscribers. The publisher/subscriber clients allow the data to be streamed through for any clients which subscribe to the specific data sources. Message gateways can also be attached to the pub/sub clients for source-specific data analysis before sending to the messaging hub.

Each anomaly detector can subscribe to one or more data sources, and publish alerts for other anomaly detectors. The incoming data are also sent into an operational database. An ETL (extract-transform-load) module can periodically pull the data from the operational database into a data warehouse for historical analysis. The analysis sometimes is conducted with the help from a GIS server that stores the geographic information. Additional statistical analysis can be performed on the historical data by an analysis server. The public health personnel can access alerts published in real time by the anomaly detectors as well as use the analysis server and data warehouse to analyze the historical data.

Ontology and metadata (as shown in Fig. 3) for data sources can often be used to assist in the interpretation of data at the appropriate abstraction level. In the disease classification area, ICD-9-CM (International Classification of Diseases, Ninth Revision, Clinical Modification) is the official system of assigning codes to diagnoses and procedures associated with hospital utilization in the United States. As an example, diseases of the respiratory system are coded with 460-519. New codes may be assigned to newly discovered diseases – such as Severe Acute Respiratory Syndrome - SARS (480.3) has just been added under the category of viral pneumonia (480) since 2003.

4. Problem Domain

The problem domains can be broadly classified according to the scale of impact:

Non-infectious disease: such as cancer, cardiovascular disease, dementia, and diabetes;

Infectious diseases: which include communicable diseases (such as flu and smallpox, which can spread from human to human), vector-born diseases which are carried by rodents or mosquitoes (such as West Nile encephalitis, Dengue fever, malaria, and Hantavirus Pulmonary Disease).

Note that non-infectious diseases of a population can potentially have a common root cause, such as caused by the same type of pollutant, similar work condition, etc.   Each of the problem domains can potentially be associated with the progression of symptoms – such as sore throat, running nose, coughing, fever, vomiting, dizziness, etc. Most of the diagnoses done before the confirmation of lab tests are based on apparent symptoms – and hence the term - syndromic surveillance.

The problem domain usually determines the data sources that will be meaningful for early detection as well as which detection methods can be applied. As an example, sewage data will be more useful for GI (gastrointestinal) related disorders, while pollutants (from Environmental Protection Agency - EPA) will be more useful for respiratory diseases or disorders.

5. Data Sources

Data sources for early detection can be categorized into:

Clinically related data – the data acquired in a clinical setting such as patient records (such as the reference information model defined by HL7), insurance claims (e.g. CMS-1500 for physician visits and UB-92 for hospital visits), prescription drug claims (NCPDP), lab tests, imaging, etc.   These claim forms can be converted into an electronic format - HIPAA 837 (HIPAA stands for Health Information Portability and Accountability Act) and vice versa. As an example, CMS-1500 contains information such as patient age, gender, geography, ICD-9 diagnosis codes, CPT procedure codes, date of service, physician, location of care, and payer type. NCPDP form for prescription drug contains, in addition to the basic patient data, Rx date written, Rx date filled, NDC (form/strength), quantity dispensed, days supply, specialty, payer type, and pharmacy.

Non-traditional Data - Emerging syndromic surveillance applications also start to use non-traditional (non-clinical) data sources to observe population behavior, such as retail sales (including over-the-counter drug sales), absentee data from school and work, chief complaints from the emergency room, and 911 calls. These data sources do not necessarily have standardized formats. Additional data sources include audio/video recording of seminars (for counting cough episodes).

Auxiliary Data - Additional data sources include animal data, environmental data (ground moisture, ground temperature, and pollutants) and micro-array data. Many of these data sources are problem domain dependent.

A number of studies have been devoted to investigating various data sources, such as the text and diagnosis code of the chief complaints from emergency department [Begier 2003, Beitel 2004, Espino 2001, Greenko 2003], 911 calls [Dockrey 2002], and over-the-counter (OTC) drug sales [Goldenberg 2002].

The data sources can be further characterized by its modality:

Structured: most of the transactional records stored in a relational database belong to this category. Each field of a transaction is often well defined. This may include a portion of the clinical & patient records - such as patient name, visit date, care provider name and address, diagnosis in terms of International Classification of Disease code (or ICD-9 code).

Semi-structured: refers to data that has been tagged by mark up languages such as XML. Emerging data standards such as HL7 enable the transmission of healthcare related information in XML encoded format.

Non-structured: Patient records, lab reports, emergency room diagnosis and 911 calls could include chief complaints in free text format. Additional non-structured data include data from various instruments such as CT, PET, MRI, EKG, ultrasound and from clinical and non-clinical data such as weather maps, videos, and audio segments.

The type of data sources is often essential in determining the methodologies for detecting anomalies.

6. Detection Methods

The analytic methods are often data source dependent.   Spatial scan statistics has been most widely used, such as in [Berkom 2003].   Other methods, including linear mixed models [Kleinman 2004], space-time clustering [Koch 2001], CuSum [O'Brien 1997], time-series modeling [Reis 2003], rule-based [Wong 2003], and concept-based [Buckridge 2002], have also been investigated and compared.   A comprehensive comparison among various detection algorithms is conducted in [Buckridge 2005] as a result of the Darpa sponsored BioALIRT program.

The detection methods are used to determine the existence of anomalies in the problem domain from available data sources, sometimes with either explicit or implicit assumptions of models.   These detection methods, as shown in Fig. 2, can be categorized into:

Inductive method: Most of the statistical methods such as those based on linear regression, Kalman filter, and spatial scan all fall into this category.   Inductive methods can be further categorized as supervised and unsupervised methods. Supervised methods usually involve learning and/or training a model based on historical data.   Unsupervised methods, on the other hand, do not require historical data for training. Unsupervised methods are more preferable for those rare events (such as those potentially caused by bio-terrorism) when there are insufficient historical data. It has been concluded in [Buckridge 2005] that spatial and other covariate information from disparate sources could improve the timeliness of outbreak detection.   Some of the existing syndromic surveillance deployments such as the CDC BioSense and New York City have leveraged both CuSum and Spatial Scan statistics.

Deductive methods: Deductive methods usually involve rule-based reasoning and inferencing.   This methodology is often used in the area of knowledge-based surveillance in which existing pertinent knowledge and data are applied in making inferences from the newly arrived data.

Hybrid inductive/deductive methods: Both inductive and deductive methods can be used simultaneously on a given problem.   Deductive methods can be used to select the proper inductive method given the prior knowledge in the problem domain and available data sources.   This often results in improved performance as compared to using either inductive or deductive methods alone [Buckridge 2005].

7. Summary and Conclusion

Early warning systems for environmental and public health applications have received substantial attention during the recent past due to increased awareness of infectious diseases and availability of surveillance technologies. In this chapter, we surveyed a number of existing early warning systems in terms of their problem domains (non-infectious disease vs. infectious disease), data sources (clinical data, non-traditional data and auxiliary data), and detection approaches (inductive, deductive, and hybrid). The rapid progress in the syndromic surveillance area seems to suggest that early detection of certain infectious diseases is feasible. Nevertheless, there is substantial interdependency among problem domains, data sources, and detection methods for syndromic surveillance. A validated early warning or anomaly detection approach for a given problem area often cannot be generalized to other problem areas due to the nature of data sources. Further studies in this area will likely to be based on the ontology (or taxonomy) of the diseases, data sources, and detection approaches.

Acknowledgment

The authors would like to acknowledge the EpiSPIRE bio-surveillance team at IBM T. J. Watson Research Center. This research is sponsored in part by the Defense Advanced Research Projects Agency and managed by Air Force Research Laboratory under contract F30602-01-C-0184 and NASA/IBM CAN NCC5-305.

The views and conclusions contained in this document are those of the authors and should not be interpreted as necessarily representing the official policies, either expressed or implied of the Defense Advanced Research Projects Agency, Air Force Research Lab, NASA, or the United States Government.

References

Begier, E. M. , D. Sockwell, L. M. Branch, J. O. Davies-Cole, L. H. Jones, L. Edwards, J. A. Casani, D. Blythe. (2003) The National Capitol Region's emergency department syndromic surveillance system: do chief complaint and discharge diagnosis yield different results? *Emerging Infectious Disease*, vol. 9, no. 3, pp. 393-396, March.

Beitel, A. J. , K. L. Olson, B. Y. Reis, K. D. Mandl. (2004) "Use of Emergency Department Chief Complaint and Diagnostic Codes for Identifying Repiratory Illness in a Pediatric Population," Pediatric Emergency Care, vol 20, no. 6, pp. 355-360, June, 2004.

Buckeridge, D. L. , J. K. Graham, M. J. O'Connor, M. K. Choy, S. W. Tu, M. A. Musen. (2002) Knowledge-based bioterrorism surveillance, *Proceedings AMIA Symposium* 2002: 76-80.

Buckeridge, D. L., H. Burkom, M. Campbell, W. R. Hoganb, A. W. Moore. [2005] Algorithms for Rapid Outbreak Detection: A Research Synthesis. *Biomedical Informatics*, 2005 April, 38(2):99-113.

Buehler, J. W., R. L. Berkelman, D. M. Hartley, C. J. Peters (2003) "Syndromic Surveillance and Bioterrorism-related epidemics," *Emerging Infectious Diseases*, 2003 Oct. vol. 9 no. 10, pp 1197-1204.

Burkom, H. S. (2003) "Biosurveillance applying scan statistics with multiple, disparate data source," Journal of Urban Health, 2003, June; 80 Suppl 1:i57-i65.

Denninghoff, Kurt R., M. H. Smith, L. Hillman. (2000) Retinal Imaging Techniques in Diabetes, *Diabetes Technology & Therapeutics*, vol. 2, no. 1, pp. 111-113.

Dockrey, M. R. , L. J. Trigg, W. B. Lober. (2002) An Information Systems for 911 Dispatch Monitoring System and Analysis, *Proceeding of the AMIA 2002 Annual Sympoisum* pp. 1008.

Duchin, J. S. (2003) Epidemiological Response to Syndromic Surveillance Signals, *Journal of Urban Health*, 2003 80:i115-i116.

Espino, J. U. , M. M. Wagner. (2001) Accuracy of ICD-9-coded chief complaints and diagnoses for the detection of acute respiratory illness, *Proceedings AMIA Symposium 2001*, pp164-168.

Glass, GE, T. L. Yates, J. B. Fine, T. M. Shields, J. B. Kendall, A. G. Hope, C. A. Parmenter, C.J. Peters, T. G. Ksiazek, C.-S. Li, J. A. Patz and J. N. Mills. (2002) Satellite imagery characterizes local animal reservoir populations of Sin Nombre virus in the southwestern United States, *Proc. National Academy of Science* 99:16817-16822.

Glass, G. E. (2001a) Public health applications of near real time weather data, *Proc. 6th Earth Sciences Information Partnership Conf.*

Glass G. E. (2001b) Hantaviruses - Climate Impacts and Integrated Assessment, *Energy Modeling Forum.*

Goldenberg, A., G. Shmueli, R. A. Caruana, S. E. Fienberg. (2002) "Early Statistical Detection of Anthrax outbreaks by tracking over-the-counter medication sales," Proceedings of the National Academy of Sciences of the United States of America, vol. 99, no. 8, pp. 5237-5240, April 16, 2002.

Goodwin, T. , and E. Noji, (2004) Syndromic Surveillance, *European Journal of Emergency Medicine*, 2004 Feb; vol. 11, no. 1, pp 1-2.

Greenko, J. , F. Mosgtashari, A. Fine, M. Layton. (2003) Clinical Evaluation of the Emergency Medical Services (EMS) Ambulance Dispatch-Based Syndromic Surveillance System, New York City, *Journal of Urban Health*, 2003 Un; 80 Suppl 1:i50-i56.

Klein, S. L., A. L. Marson, A. L. Scott, GE Glass. (2001a) Sex differences in hantavirus infection are altered by neonatal hormone manipulation in Norway rats, *Soc Neuroscience.*

Klein, S. L. , A. L. Scott, G. E. Glass. (2001b) Sex differences in hantavirus infection: interactions among hormones, genes, and immunity, *Am Physiol Soc.*

Kleinman, K., R. Lazarus, R. Platt. (2004) "A generalized linear mixed models approach for detecting incident clusters of disease in small areas, with an application to biological terrorism," *American Journal of Epidemiology*, 2004 Feb 1; 159(30: 217-224.

Koch, M. W. , S. A. Mckenna. (2001) "Near Real Time Surveillance Against Bioterror Attack Using Space-Time Clustering, Technical Report, (http://www.prod.sandia.gov/cgi-bin/techlib/access-control.pl/2001/010820p.pdf)

Mostashari, F. , J. Hartman. (2003) "Syndromic Surveillance: A Local Perspective," *Journal of Urban Health*, 2003; 80 Suppl 1:i1-i7.

O'Brien, S. J., P. Christie. (1997) Do CuSums have a role in routine communicable disease surveillance, *Public Health* , July, 11194: 255-8.

Pavlin, J. A. (2003) Investigation of Disease Outbreaks Detected by Syndromic Surveillance Systems, *Journal of Urban Health*, 2003; 80:i107-i114.

Sosin, D. M. (2003) Syndromic Surveillance: The case for skillful investment. Biosecurity and Bioterrorism: Biodefense strategy, *Practice and Science* 2003 vol 1, no. 4, pp.247-253.

Stoto, M. A., M. Schonlau, and L. T. Mariano. (2004) Syndromic Surveillance: Is It Worth the Effort? Chance, Vol. 17, No. 1, 2004, pp. 19-24

Reis, B. Y. and K. D. Mandle. (2003) Time Series Modeling for Syndromic Surveillance, *BMC Medical Informatics and Decision Making*, Jan 23; 3(1); 2.

Wagner, M. M. , F. C. Tsui, J. U. Espino, V. M. Dato, D. F. Sittig, R. A. Caruana, L. F. McGinnis, D. W. Deerfield, M. J. Druzdzel, D. B. Fridsma. (2001) The emerging science of very early detection of disease outbreaks, *Journal of Public Health Management Practice*, 2001 Nov; vol 7, no. 6, pp. 51-59.

Wong, W. , A. Moore, G. Cooper, and M. Wagner. (2003) "Rule-Based Anomaly Pattern Detection for Detecting Disease Outbreaks," *Journal of Urban health*, June 80 Suppl 1: i66-i75.

CHAPTER 2

TIME-LAPSE CELL CYCLE QUANTITATIVE DATA ANALYSIS USING GAUSSIAN MIXTURE MODELS

Xiaobo Zhou[1,2], Xiaowei Chen[1,2], Kuang-Yu Liu[1,2],
Randy King[3], and Stephen TC Wong[1,2]

[1]*Harvard Center for Neurodegeneration and Repair-Center for Bioinformatics, Harvard Medical School*

[2]*Functional and Molecular Center, Radiology Department, Brigham & Women's Hospital*

[3]*Department of Cell Biology, and Institute of Chemistry and Cell Biology, Harvard Medical School*

Abstract

Time-lapse fluorescence microscopy imaging provides an important method to study the cell cycle process under different conditions of perturbation. Existing methods, however, are rather limited in dealing with such time-lapse data sets, while manual analysis is unreasonably time-consuming. This chapter presents statistical data analysis issues and statistical pattern recognition to fill this gap. We propose to apply Gaussian mixture model (GMM) to study the classification problems. We first propose to model the time-lapse cell trace data by using auto-regression (AR) model and to filter the cell features using this model. We then study whether there is significant difference in cell morphology between untreated and treated cases using Pearson correlation and GMM. Furthermore, we propose to study cell phase identification using GMM, and compare with other traditional classifiers. Once we identify the cell phase information, then we can answer questions such as when the cells are arrested. We employ the

ordered Fisher clustering algorithm to study this problem. The GMM
is shown to have a high accuracy to identify treated and untreated cell
traces. From the cell morphologic similarity analysis, we found that
there is no significant correlation between untreated and treated cases.
For cell phase identification, the experiments show the GMM has the
best recognition accuracy. Also, the experiments show the result from
the Fisher clustering is consistent with biological observations as well
as KS test.

1. Introduction

To better understand drug effects on cancer cells, it is important to
measure cell cycle progression (e.g., interphase, prophase, metaphase,
and anaphase) in individual cells as a function of time. Cell cycle
progress can be identified by measuring nucleus changes. Automated
time-lapse fluorescence microscopy imaging provides an important
method to observe and study nuclei in a dynamic fashion [12, 13, 15].
However, existing software is inadequate to extract cell phase
information of large volumes of high resolution time lapse microscopy
images. The assignment of a cell to a particular phase is currently done
by visual inspection of images [8, 16, 28]. Manual inspection is subject
to investigator bias, cannot be easily confirmed by other investigators,
and, more importantly, cannot scale up for high throughput studies, such
as drug screens and cytologic profiling. This motivates us to develop an
automated system, CellIQ (Cellular Image Quantitator) for cell phase
identification to provide advantages over current practice in speed,
objectivity, reliability, and repeatability.

   To employ such dynamic cellular data sets [4] to investigate cytologic
mechanisms, we discuss four questions arising from cell cycle
quantitative data. To answer these questions, we propose to apply
Gaussian mixture model (GMM)[21, 25, 34] to study the classification
problems. If we know the distribution of observations, the maximum
likelihood estimation [21] is an optimal method in pattern recognition.
How to get a high accuracy distribution of observations is a critical issue.
In this study, the maximum likelihood estimation of Gaussian mixture
model via Expectation Maximization (EM) algorithm [21, 25] is used to

estimate distribution of observations. Our major contributions in this study are summarized in the following four aspects:

- Cell feature preprocessing. We propose to model the time-lapse data by using auto-regression model and then identify the different traces between untreated and treated cases using GMM. Although there are many methods (for example Fast Fourier transformation and wavelet transformation, just mention a few) to filter time-series data, they are not good for pattern recognition of dynamic cell cycle sequences. Motivated by the feature extracting in speech recognition, we apply auto-regression model, also known as linear prediction coefficients (LPC) [20], to model time-series cell cycle data [4]. The linear prediction coefficients will then be employed as features of each cell trace for treated and untreated cell sequence separation (we call this trace identification). The AR model is also employed to filter the cell features.
- Cell morphologic similarity analysis. Is there any significant cell morphologic similarity between untreated and treated cases? The answer to this question can help determining whether we need to separate different cases (untreated and treated cases) or combine different cases to perform cell phase identification. We employ Pearson correlation coefficients [7] and GMM classification [25, 34] to study this problem.
- Cell phase identification. Whether the automatic cell-IQ system is successful or not is mainly dependent on whether we can identify the cell phases automatically. Since we have lots of cell data available, it is possible to estimate the distribution of the large amount of data sets. We employ Gaussian mixture model and expectation maximum (EM) [21] algorithm to get high accuracy estimation of the data distribution. We also compare the GMM method with other existing classifiers such as K nearest neighbor classifier [17], neural networks [24, 26], and decision tree [3].
- Cell cycle time-lapse cluster analysis. Can we identify the period that the cell phases change dramatically from metaphase to apoptosis? We propose to apply the ordered Fisher clustering algorithm [11] to this problem because the traditional clustering methods [31] do not consider the order property of time-series data.

The rest of this chapter is organized as follows. In Section 2, we summarize the quantitative features extracted from time-lapse microscope images by using a new image processing methods [5]. Meanwhile, we present the modeling of the cell trace data by using the auto regression model. In Section 3, we design and formulate the four issues that are arisen from the quantitative time-lapse data. We present the GMM and other classification methods in Section 4. In Section 5, experimental results are provided. Finally, Section 6 contains the discussion and conclusions.

2. Material and Feature Extraction

The data is composed of HeLa cells expressing H2B-GFB with five untreated and five treated using 100nM (nanomolar) taxol and imaged for a 5-day period, with images acquired every 15 minutes beginning 24 hours after taxol treatment. Next we summarize the imaging processing and feature calculation for time-lapse cellIQ data quantitative analysis and propose a method to extract cell trace feature by using the auto regression model.

2.1. *Material and cell feature extraction*

Nuclei segmentation, which segments individual nuclei from their background, is critical for cell phase identification. In time-lapse fluorescence microscopy images, nuclei are bright objects protruding out from a relatively uniform dark background, see an example in Fig. 1. They can be segmented by histogram thresholding. In this work, the ISODATA algorithm was used to perform image thresholding [23]. This algorithm correctly segments most isolated nuclei, but it is unable to segment touching nuclei. To handle the case of touching nuclei, a watershed algorithm was used [1, 22]. Watershed algorithm always causes nuclei be to over-segmented. To solve this problem, a size and shape based merging processing [5] is devised to merge over-segmented nuclei pieces together. We also developed a size and location based method for nuclei tracking, which can successfully deal with nuclei

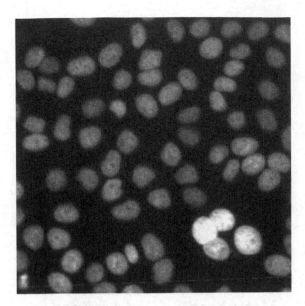

Figure 1: Gray level image of nuclei.

changes during cell mitosis and cell division [5]. In this method, a match processing is used to set up the correspondence between nuclei in the current frame and the frame following the current frame. Ambiguous correspondences are solved only after nuclei move away from each other. Nuclei division is considered as a special case of splitting. Daughter cell nuclei are found and verified by a process which matches the center of gravity of daughter cell nuclei with the center of gravities of their parents. After nucleus segmentation has been performed, it is necessary to perform a morphological closing process on the resulting binary images in order to smooth the nuclei boundaries and fill holes insides nuclei [4]. These binary images are then used as a mask on the original image to arrive at the final segmentation. From this resulting image, features can be extracted. Figure 2 shows nucleus appearances in different phases.

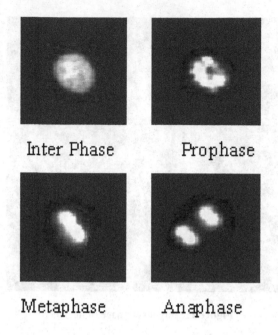

Inter Phase   Prophase

Metaphase   Anaphase

Figure 2: Nucleus appearances in different phases.

Once the individual nuclei have been detected by the segmentation procedure, it is then necessary to identify its cell phase. Cell phase are identified by comparing nuclei shape and intensity. To identify the shape and intensity differences between different cell phases, a set of twelve features were extracted based on the experience of expert cell biologists. We calculate the following features for each cell [5]: (1) the maximum of the intensity, (2) the minimum of the intensity, (3) the average of the intensity, (4) the standard deviation of gray level within a cell nucleus, (5) major axis-the length of a cell nucleus, (6) minor axis-the breadth of a cell nucleus, (7) elongation-the ratio major axis to minor axis, (8) area-number of pixels within a cell nucleus, (9) perimeter-the total length of edges of a cell nucleus, (10) compactness-the ratio perimeter to Area, (11) convex perimeter-perimeter of the convex hull, and (12) roughness-the ratio convex perimeter to perimeter. For cell trace feature extraction, we next present the auto-regression model to deal with the time-series cell trace signal. The coefficients of this model will be employed as trace features.

2.2. Model the time-lapse data using AR model

Since numerous of cells are either dead, lost in tracking, or disappeared after several days, it is hard to classify the different traces by using the original cell shape features because each trace has a different number of features or dimensions. Another problem is that there exists strong noise in the features due to imperfect segmentation. So robust modeling the shape change critical. Here we model the time-lapse data using an auto-regression model. We assume the current input sample $y(n)$ is approximated by a linear combination of the past samples of the input signal. In our study, the $\{y(n)\}_{n=1}^{N}$ represent the N time points of one cell feature. The prediction of $y(n)$ is computed using a finite impulse response filter by:

$$\tilde{y}(n) = \sum_{k=1}^{p} a(k)y(n-k) \tag{1}$$

where p is the prediction order and $a(k)$ is the linear prediction coefficients (LPC) which can be solved by minimizing the sum of the squares of the errors $e(n) = y(n) - \tilde{y}(n)$. The autocorrelation method minimizes the prediction error $e(n)$. For a block of length N, define $R(i) = p^{-1} \sum_{n=i}^{p-1} u(n)u(n-i)$, where $u(n) = y(n)w(n)$ is a windowed version of the considered block $y(n)$, $n = 0, ..., p-1$. Normally a Hamming window [20] is used design $w(n)$. LPC uses the Levinson-Durbin recursion [20] to solve the following normal equations $\sum_{k=1}^{p-1} a(k)R(i-k) = R(i)$, $i = 1, ..., p$. For the detailed recursive parameter estimation algorithm, we refer to [20]. The computation of the linear prediction coefficients is often referred to as the autocorrelation method. Another advantage of this model is that we do not need to perform any special normalization processing to the coefficients for trace identification. Assume there are total time points, the coefficient of determination [10]

$$r = 1 - \frac{\sum_{i=1}^{N}\left(y(i) - \tilde{y}(i)\right)^2}{\sum_{i=1}^{N}\left(y(i) - \overline{y}\right)^2} \qquad (2)$$

is usually used for measuring the model fitting accuracy, where $\overline{y}$ is the mean of the signal $\{y(n)\}_{n=1}^{N}$. Since $a(1)$ is always 1, the $v-1$ coefficients of $\{a(i)\}_{i=2}^{v}$ will be employed as features of each trace associate with one cell feature. In this study, we set $v = 10$ and $p = 20$. The filtered cell features $\tilde{y}(n)$ and the LPC coefficients are then employed for further analysis.

3. Problem Statement and Formulation

The data is composed of HeLa cells expressing H2B-GFB with five untreated and five treated using 100nM (nanomolar) taxol, and imaged for 5 day period with images acquired 15 minutes beginning 24 hours after taxol treatment. Twelve features are extracted for each cell from the Hela cells expressing H2B-GFP with untreated and treated images. For each cell, we have n features as introduced in the last section. Denote the jth feature of ith cell as x_{ij}, where $1 \leq j \leq n$ and $1 \leq i \leq m$, m is the total number of samples such as the number of cells or number of traces. Next we will define four different questions arising from time-lapse cell cycle quantitative data.

Q1. Trace identification. Since we imaged for a 5-day period with images acquired every 15 minutes, we totally have 480 time points for each cell trace. After using the AR model to model the each trace, and we extract $v-1$ coefficients for each cell feature with 480 time points. We then totally have $12(v-1)$ coefficients (or features) for each trace. In this way, each trace has the same number of features. We still use n to denote the number of features and x_{ij} to denote trace features, where $1 \leq j \leq n$ and $1 \leq i \leq m$ and m denotes the total number of traces. Denote I_0 and I_1 as the index sets of untreated and treated cell traces, respectively. Then the two classes $\mathbf{X}_k = \{x_{ij}, i \in I_k, 1 \leq j \leq n\}$ are the untreated and treated cell trace feature sets, where $k = 0,1$. We then can employ Gaussian mixture model to model each class and classify the two classes.

Q2. Cell morphologic similarity analysis. We combine all cells with four different phases, namely, interphase cells, prophase cells, metaphase cells, and anaphase cells. Without loss of generality, we consider the case of interphase. The cell data is first normalized by z-score [4]. Note that we discuss the cell characteristics, hence we use x_{ij} to denote the jth feature of ith cell, where $1 \leq j \leq n$. Assume there are m_1 and m_2 untreated and treated cells in interphase, and we have totally two classes in this phase. For the cell morphologic similarity analysis, we will first calculate Pearson correlation coefficients between the untreated and treated cells. Furthermore, we again employ Gaussian mixture model to model each class and classify the two classes to see if we can separate them.

Q3. Cell cycle phase identification. We perform HeLa cell phase identification using pattern recognition. The cell data is again first normalized by z-score. We still denote x_{ij} as the jth feature of ith cell, where $1 \leq j \leq n$. Obviously we have four classes, namely m_1 interphase cells, m_2 prophase cells, m_3 metaphase cells, and m_4 anaphase cells. Considering there exist different kinds of classifiers, we will compare the GMM with KNN, neural networks and decision tree.

Q4. Cell cycle time-lapse cluster analysis. The goal is to identify in which period the cell phases change dramatically. For the orderly, time-series cell sequence data, we will adopt an ordered clustering method named Fisher clustering algorithm. It differs from traditional clustering methods because traditional clustering does not consider the ordering of data. We will briefly introduce the principle of Fisher clustering algorithm in the next section.

The first three questions can be categorized into classification problems. We will employ the optimal Bayesian classifier or maximum likelihood classifier to study them. Here we briefly discuss the formulation of the optimal classifier. Suppose that an object which is described by a feature vector $\mathbf{x}$ belongs to one of the M classes: $C_1, C_2, ..., C_M$. The optimal Bayesian decision rule for classifying $\mathbf{x}$ is given by

$$
\begin{aligned}
\hat{C}(\mathbf{x}) &= \underset{1 \le K \le M}{\arg\max} \, P\big(C(\mathbf{x}) = k \mid \mathbf{x}\big) \\
&= \underset{1 \le K \le M}{\arg\max} \, P\big(\mathbf{x} \mid C_k\big) P(C_k).
\end{aligned}
\tag{3}
$$

We also call it *maximum a posteriori* estimation. For simplification, we assign equal priori probability $P(C_k)$ to each class, hence we can ignore this term. Then the main objective is to estimate the conditional probability distribution of the feature vector:

$$
P\big(\mathbf{x} \mid C_k\big), \quad k = 1, 2, ..., M
\tag{4}
$$

based on the training data $\big\{(\mathbf{x}_i, C(\mathbf{x}_i))\big\}_{i=1}^{m}$ by using Gaussian mixture model. Denote the set of feature vectors used for training as $\mathbf{X} = \{\mathbf{x}_1, \mathbf{x}_2, ..., \mathbf{x}_m\}$. For each class C_k, denote $\mathbf{X}_k = \{\mathbf{x} \in \mathbf{X} : C(\mathbf{x}) = k\}$ as the set of training samples associated with this class. The details of estimating (4) are given in the next section.

4. Classification Methods

In this section, we first briefly describe how to employ expectation maximization (EM) algorithm to estimate the conditional probability distribution in (4) which is defined as a Gaussian mixture model. In order to compare with other existing classifiers, we also briefly introduce other three classifiers including KNN classifier, NN classifier and decision tree. Meanwhile we briefly introduce the principle of the ordered Fisher clustering algorithm.

4.1. *Gaussian mixture models and the EM algorithm*

The expectation maximization (EM) algorithm is an iterative method for approximating the ML estimates of the parameters in a mixture model [21, 25, 34]. Alternatively, it can be viewed as an estimation problem involving incomplete data in which each unlabeled observation in the mixture is regarded as missing its label. Under the mixture model, the distribution of the data $\mathbf{x} \in \mathbb{R}^n$ is given as:

$$f(\mathbf{x}, \Phi) = \sum_{i=1}^{K} w_i f_i(\mathbf{x}; \phi_i)$$

where w_i is the mixing proportion, f_i is the component density parameterized by ϕ_i, and K is the total number of components. The mixture density f is then parameterized by $\Phi = (\phi_1, \phi_2, ..., \phi_K)$.

Under the incomplete data formulation, each unlabeled sample $\mathbf{x}$ is considered as the labeled sample $\mathbf{y}$ with its class origin missing. Let $g(\mathbf{x} | \Phi)$ be the probability density function (pdf) of the incomplete data $\mathbf{x}$ and let $f(\mathbf{y} | \Phi)$ be the pdf of the completely labeled data $\mathbf{y}$. The maximum likelihood (ML) estimation then involves the maximization of the log likelihood of the incomplete data $\log g(\mathbf{x} | \Phi)$, namely, the ML estimation is given by $\hat{\Phi} = \arg\max_{\Phi} g(\mathbf{x} | \Phi)$. The estimation is complicated by the fact that the sample origin is missing. Hence, the EM algorithm uses the relationship between $f(\mathbf{y} | \Phi)$ and $g(\mathbf{y} | \Phi)$. Using an iterative approach, the EM algorithm obtains the ML estimator by starting with initial estimate $\Phi^{(0)}$ and repeating the following two steps at each iteration:

- Expectation step: Determine

$$(Q(\Phi | \Phi^{(m-1)}) = \mathbf{E}\left\{ f(\mathbf{y} | \Phi) | \mathbf{x}, \Phi^{(m-1)} \right\}$$

- Maximization Step: Choose

$$\Phi^{(m)} = \arg\max_{\Phi} Q(\Phi | \Phi^{(m-1)})$$

where m is the iteration number. The most general representation of the pdf, for which a re-estimation procedure has been formulated, is a finite mixture of the form. In the Gaussian mixture model, the continuous density function in (4) is defined as the following Gaussian mixture model:

$$P\left(\mathbf{x} \mid C_i\right) = \sum_{k=1}^{K} w_k \mathbb{N}(\mathbf{x}; \boldsymbol{\mu}_k, \Sigma_k) \tag{5}$$

for the ith class, where w_k is the mixture coefficient for the kth mixture $\mathbb{N}(\mathbf{x}; \boldsymbol{\mu}_k, \Sigma_k)$ is Gaussian distribution with mean vector $\boldsymbol{\mu}_k = (\mu_{k1}, \mu_{k2}, ..., \mu_{kn})$ and covariance matrix Σ_k for the kth mixture component. The mixture number K is determined as $K = 2^{\lfloor \log_{10} m + 1 \rfloor}$ by experience, where m is the total number of samples in class i. The mixture coefficients w_k satisfy the stochastic constraint $\sum_{k=1}^{K} w_k = 1$, $w_k \geq 0$. Using the above EM algorithm, the re-estimation formulas for the parameters of the mixture density in (5) are given in Appendix A.

4.2. K-Nearest Neighbor (KNN) classifier

A classifier can provide a criterion to evaluate the discrimination power of the features for the feature subset selection. A KNN classifier is chosen for its simplicity and flexibility [5, 17]. Each cell is represented as a vector in an n-dimension feature space. The distance between two cells is defined by the Euclidean distance. A training set is used to determine the class of a previously unseen nucleus. The classifier calculated the distances between an unseen nucleus and all nuclei in the training set. Next, the K cells in the training set which are the closest to the unseen nucleus are selected. The phase of this cell is determined to be the phase of the most common cell type in the K nearest neighbors.

4.3. Neural networks

The most widely used neural network classifier today is Multilayer Perceptron (MLP) network [24] which has also been extensively analyzed and for which many learning algorithms have been developed. The principle of the network is that when data from an input pattern is presented at the input layer the network nodes perform calculations in the successive layers until an output value is computed at each of the output nodes. The multilayer feed-forward neural network used is a three layer network with one hidden layer of ten units. There are twelve units in the input layer corresponding to the feature vector and two units in the output layer. A sigmoid activation function is frequently used in the output layer. The training usually adopts backward propagation algorithm. In the cell phase identification, this traditional neural network does not work, but the general regression neural network does [26]. Generalized regression neural network is a kind of radial basis network that is often used for function approximation.

4.4. *Decision tree*

Conventional decision tree algorithms, such as the parallelepiped classifier, use one feature at a time and search for optimal axes parallel separating planes at each node of decision tree. Here we adopted a piecewise linear decision classifier [3] to better approximate complex class boundaries. Associated with each decision node in the binary tree is hyperplane that divides the multidimensional space into two separate regions. A genetic algorithm is used to search for optimal decision boundaries using training data. Node splitting continues until one of several stopping conditions is satisfied, such as the minimum terminal node size. The essence of the binary decision tree classifier is to split the training data at each node into two subsets, so that the data after splitting (i.e., in each subtree) becomes purer, using a suitable measure of call homogeneity, in comparison to the parent node. Let $\mathbf{X}_t = \{\mathbf{x}_1, \mathbf{x}_2, ..., \mathbf{x}_{N_t}\}$ be the training data arriving at the current node t with each sample belonging to one of M classes. The ith observation, $\mathbf{x}_i = [x_{i1}, x_{i2}, ..., x_{im}]^T$, in the training data with d features, can be assigned to one of two subsets $\mathbf{X}_{tL}(w)$ and $\mathbf{X}_{tR}(w)$, for the left and right children nodes respectively, using the linear decision function

$$\mathbf{x}_i^T \mathbf{w} + w_0 \le 0 \qquad (6)$$

Denoting the impurity of node t as $I(\mathbf{X}_t)$, the impurity reduction gained by splitting node t is calculated by subtracting the weighted average impurities and the Twoing rule [3] is employed to calculate $I(\mathbf{X}_t)$, $I(\mathbf{X}_{tL})$, $I(\mathbf{X}_{tR})$. The objective is to find a weight vector

$\mathbf{w} = [w_1, w_2, ..., w_n]^T$ that maximizes the impurity reduction for a given node split. The optimal decision function for each node is given by using a genetic algorithm [6, 9] which provides a global optimization procedure. In our experiment, the minimum terminal node size are 300, impurity reduction is criterion is 0.001. GA population is 120, GA generation 20, GA minimum generation 10, GA Crossover probability 0.9 and GA Mutation probability 0.01.

4.5. *Fisher clustering*

In order to introduce the Fisher clustering algorithm, we need to first define within-class mean variance as a within-class diameter. Denote the cell phase information z_{kl} as the phase of the lth trace and kth cell or kth time point, where $1 \le l \le m$ and $1 \le k \le n$. Here m denotes the total number of traces and n is 480 in this study. Since the time-lapse data is time series data, we cannot change the order of the time points. Here the within-class diameter $D(i, j)$ is defined as:

$$D(i, j) = \frac{1}{j-i+1} \sum_{l=i}^{j} \sum_{k=1}^{n} (z_{lk} - \overline{z}_l)^2, 1 \le i \le j \le n,$$

where $\overline{z}_l$ is the mean of z_{lk} for $k = i, ..., j$. The Fisher clustering algorithm is a dynamic optimal algorithm to find the optimal partition points without changing the ordering of the data. The algorithm is summarized in Appendix B. For more details about the discussion of this algorithm, we refer to [11, 29]. Since we want to see in which period the cell phase in the time-lapse cellIQ data is dramatically changed or not, it is critical to keep the ordering of the data. Traditional clustering methods [31] are not able to accomplish this aim.

5. **Experimental Results**

We performed experiments to answer the four questions arisen in time-lapse data analysis. They are trace modeling and trace identification, cell morphologic similarity analysis, cell phase identification, and cluster analysis of time-lapse cell phase information.

5.1. *Trace identification*

As discussed before, since certain cells are either dead, lost tracking, or simply disappeared after several days of continued imaging, it is hard to classify the different traces by using the original features. Another problem is that there is strong noise embedded in the features. So modeling the shape change is important. We first model the time lapse data and then apply the four classifiers to identify the different traces. In Fig. 3, we show that some examples of modeling time-series features

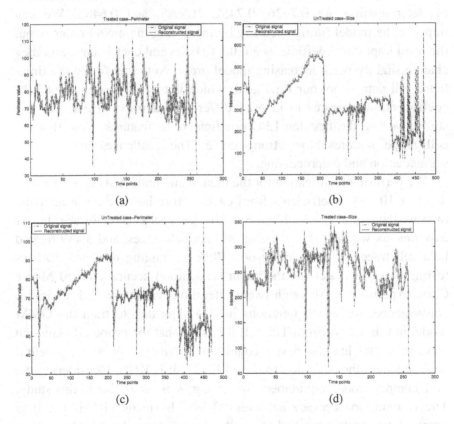

(a)   (b)

(c)   (d)

Figure 3: Model one time-series feature using LPC method, the original one, and the fitted one: (a) an example of modeling of the size (area) variable as a cell trace with 480 time points for untreated case, (b) an example of modeling of the perimeter variable as a cell trace with 480 time points for untreated case; (c) an example of modeling of the size variable as a cell trace with 260 time points for treated case; and (d) an example of modeling of the perimeter variable as a cell trace with 260 time points for treated case.

using LPC method, the original one, and the fitted one: (a) an example of modeling of the size (area) variable as a cell trace with 480 time points for untreated case, (b) an example of modeling of the perimeter variable as a cell trace with 480 time points for untreated case; (c) an example of modeling of the size variable as a cell trace with 260 time points for treated case; and (d) an example of modeling of the perimeter variable as a cell trace with 260 time points for treated case. Obviously, the signals are changed drastically with time, but it is shown that the proposed LPC model can still fit the original signal well. The corresponding coefficients of determination are 0.7926, 0.7452, 0.6688, and 0.6483. We can improve the model fitting accuracy by increasing the model order p, but the most important coefficients are the first several coefficients, and they change slightly when increasing model order. Note that fitting the time-lapse cell data is not our final goal, which is to use the important LPC coefficients as features to identify different cell traces. Hence, in this study, we use the first ten LPC coefficients as features. Note that the cell shape features have strong noise. The challenges are accurate segmentation and preprocessing.

We perform classification for the treated and untreated cases. We take the first 10 LPC coefficients from each feature based time-series data, then we have total (10-1)*12=108 coefficients for each cell trace. In our experiments we have 2,073 untreated hela cell traces, and 3,439 treated hela cell traces. We randomly take 70% as training data and 30% as testing data, and calculate the mean recognition accuracy of 50 Monte Carlo experiments. Although twenty iterations can make EM algorithm convergence, we set 50 iterations in EM algorithm to train the GMM model in this study. From Table 1, it is seen that the proposed Gaussian mixture model has the best recognition accuracy, followed by neural networks, and then decision tree, and finally it is KNN. Considering the high computational requirement for testing, K is set as five in this study. The recognition accuracy achieves 93.14% by using GMM. So, it is practical to use this method in production systems. It should be more useful when there are different cell types and drugs of different concentration.

Table 1: Trace recognition accuracy by using the four classifiers.

	GMM	KNN	NN	Decision Tree
Untreated	92.37	88.49	91.17	90.20
Treated	93.84	87.30	92.32	93.15
Total	93.14	87.74	91.87	91.46

5.2. *Cell morphologic similarity analysis*

We combine all cells with four different phases, namely, interphase cells, prophase cells, metaphase cells, and anaphase cells, but ignore the time period, then perform differential feature discovery and HeLa cells classification for different phases. Figure 4 shows Pearson correlation coefficients between the same phase with untreated and treated cases. The x-axis is the number of cells in untreated case, and y-axis is the mean of the absolute value of the Pearson correlation between a cell in untreated case and every cell in treated case. Figure 4(a): the Pearson correlation coefficients for phase 1-inter phase, (b) for phase 2-prophase, (c) for phase 3-metaphase, and (d) for phase 4-anaphase. From these figures, it is seen that the Pearson correlation coefficients are very low. That means there is no obvious correlation between the cell morphology in untreated case and the cell morphology in treated case. Let us take a close look at the Pearson correlation coefficients: In Fig. 4(a), (b), (c) and (d), the maximum correlation coefficients are 0.5467, 0.1622, 0.5428, and 0.4100; and the minimum correlation coefficient are 0.0696, 0.0324, 0.1037, and 0.0429; as well as the mean of correlation coefficients are 0.4276, 0.1248, 0.4123 and 0.3228. All of the correlation coefficients are shown to be of low values.

We perform classification for the four different phases. In this experiment we take 10,000 untreated hela cell lines, and 10,000 treated hela cell lines. We randomly take 50% as training data, 50% as testing data, and then calculate the mean recognition accuracy of 50 Monte Carlo experiments. The recognition accuracy is shown in Table 2.

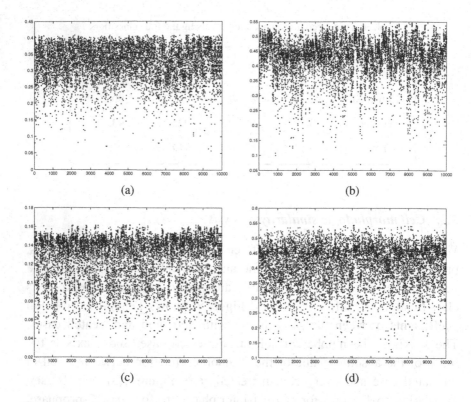

Figure 4: Pearson correlation coefficients between untreated and treated cases with the same phase. The x-axis is the number of cells, and y-axis is the absolute value of the Pearson correlation coefficients. (a): the Pearson correlation coefficients for phase 1-inter phase, (b): the Pearson correlation coefficients for phase 2-prophase, (c): the Pearson correlation coefficients for phase 3-metaphase, and (d): the Pearson correlation coefficients for phase 4-anaphase.

Firstly, the EM algorithm outperforms the KNN and decision tree classifiers. Secondly, the Gaussian mixture model and neural networks have similar high accuracy (the total recognition accuracy is between 98% and 99%) to classify the treated and untreated cells for the four phases. That means that the cells in untreated and treated cases are well-separated. This result confirms that the result obtained from the Pearson correlation coefficients in Fig. 4, namely, there is no obvious correlation between the cell morphology in untreated case and the cell

Table 2: Cell recognition accuracy by using the four classifiers.

Phase 1	GMM	KNN	NN	Decision Tree
Untreated	99.71	86.06	99.58	96.37
Treated	99.74	92.98	99.47	98.25
Total	99.72	89.50	99.49	97.82
Phase 2				
Untreated	99.16	96.47	99.17	96.33
Treated	98.12	90.07	94.57	94.86
Total	98.67	95.58	98.23	98.82
Phase 3				
Untreated	98.69	88.47	98.30	94.89
Treated	98.21	94.04	98.49	98.27
Total	98.41	91.28	98.40	96.13
Phase 4				
Untreated	99.17	92.74	99.22	99.79
Treated	97.72	90.08	96.93	92.57
Total	98.78	91.83	98.42	96.31

morphology in treated case. Hence we need to treat cell phase identification in untreated case and treated case separately. This also tells us that, in order to identify cell phase effectively, we need to train different models for different cells under various treatment conditions.

5.3. *Phase identification*

Cell phase identification is the key problem in time-lapse cell automation system and cell-cycle data analysis. In [5], we already discussed this problem in a small data set. Here we compare the performance of the four classifiers for cell phase identification using 10,000 cell lines for each phase. If the number of cells in some phase is less than 10,000, we use all cells in that phase for cell phase identification. We randomly take 50% as training data, 50% as testing data, and calculate the mean recognition accuracy of 50 Monte Carlo experiments. We do not consider any special rules and only study the general property of the four phases to see if the cell phase can be identified automatically without context information. We found that the traditional two-layer neural network did not work (the recognition accuracy only has about 50%). One reason may be that the data sets have a highly nonlinear property because we can see the K nearest neighbor classifier works very well. The generalized regression neural network with Gaussian kernel [26], on the other hand, performs better than the two-layer neural networks in cell phase identification. So here we compare the Gaussian, mixture model, K nearest neighbor classifier, regression neural networks, and decision tree. Table 3 shows that cell phase identification accuracy by using the four classifiers for the untreated case. Table 4 shows cell phase identification accuracy by using the four classifiers for the treated case. From the two tables, both GMM and KNN show similar recognition accuracy. They both perform better than neural networks and decision tree. In the untreated case, the recognition accuracy of GMM and KNN are 84.21% and 83.28%, and in the treated case, the recognition accuracy of GMM and KNN are 88.71% and 88.12%. It is also seen that all classifiers have poor recognition accuracy to identify phase 3 of both untreated and treat cases. One reason is that cells in phase 2 and phase 3 share similar morphology.

Table 3: Cell phase recognition accuracy by using the four classifiers for untreated case.

	GMM	KNN	Regression NN	Decision Tree
Phase 1	92.03	90.74	63.11	91.54
Phase 2	90.75	92.19	92.62	54.01
Phase 3	64.10	69.22	68.20	34.75
Phase 4	95.72	94.36	65.94	95.43
Total	84.21	83.28	72.25	73.43

Table 4: Cell phase recognition accuracy by using the four classifiers for treated case.

	GMM	KNN	Regression NN	Decision Tree
Phase 1	94.03	90.70	66.13	92.37
Phase 2	89.52	95.09	93.58	64.21
Phase 3	83.21	81.00	71.28	50.38
Phase 4	96.24	97.62	64.79	97.19
Total	88.71	88.12	73.82	75.63

5.4. *Cluster analysis of time-lapse data*

In this experiment, HeLa cells expressing H2B-GFP were either untreated, or treated 100 nM (nanomolar) taxol, and imaged for a 5 day period (with images acquired every 15 minutes) beginning 24 hours after taxol treatment. Figure 5(a) shows the manual analysis of a single field of cells that were not treated with taxol. Individual nuclei were identified in the first frame of the movie, and then followed throughout the time lapse experiment. Timing of the division of each cell was plotted as a function of time using a color code that designated when each cell entered prophase, prometaphase/metaphase, or anaphase/telophase (orange). Normal interphase nuclei are illustrated in white, whereas abnormal interphase nuclei are shown in grey. Cells that underwent apoptosis are indicated in black. Each row across the figure indicates the division history of a particular cell.

Figure 5(a) also illustrates that most cells spend a brief period in mitosis (the green bars are not very long). We found that many cells have divided a second or third time during the experiment and undergone divisions synchronously, that is why the bars appear thicker in later divisions. At the latter part of the cell cycle movie, few cells entered mitosis due to contact inhibition in the confluent monolayer. A small proportion of untreated cells exhibited abnormal, prolonged mitosis (illustrated by long green bars, indicative of mitotic arrest). Other cells

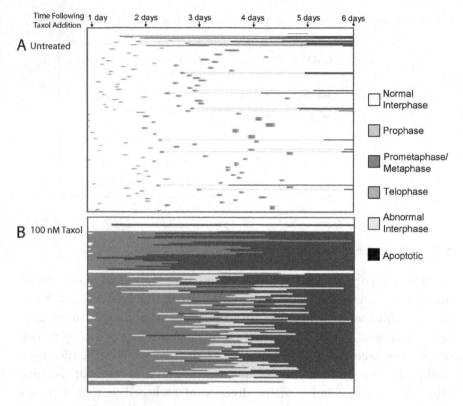

Figure 5: (a) untreated cell phase figure, and (b) treated cell phase figure (courtesy Dr. Susan Lyman, Department of Cell Biology, Harvard Medical School).

underwent mitosis and then formed an abnormal interphase nucleus (extended gray bars). These cells eventually entered apoptosis after certain delay. In contrast, cells treated with 100 nM taxol [Fig. 5(b)] showed a pronounced mitotic arrest that lasted as long as 3-4 days. Approximately 25% of cells underwent apoptosis directly from the mitotic arrest [top of Fig. 5(b)], whereas the remainder exited mitosis and formed abnormal interphase nuclei that subsequently underwent apoptosis.

We perform cluster analysis for HeLa cell time-lapse phase information for both untreated and treated cases. We apply Fisher clustering algorithm to the treated case. When the treated data is partitioned into two classes, the partition point is 192, namely, the 48th hour. We then partition them into three classes, the partition points are 192 and 410, namely, the 48th hour (2 days), and the 102th hour (4 days). According to the above biological observation, we can find the similar results, but the biological observation only can give us a feeling, does not tell us the exact partition points. For the untreated case, there exists no such special pattern, so Fisher clustering does not give any interesting discovery. We can apply other clustering technologies, such as hierarchial clustering of single linkage or other clustering of fuzzy c-means [31], to mine the data set to investigate when to begin the cell cycle.

We then run Kolmogorov–Smirnov test for the four phases. Figure 6 shows Kolmogorov–Smirnov tests for interphase, prophase, metaphase, and anaphase. The data sets are collected from the first three days. There

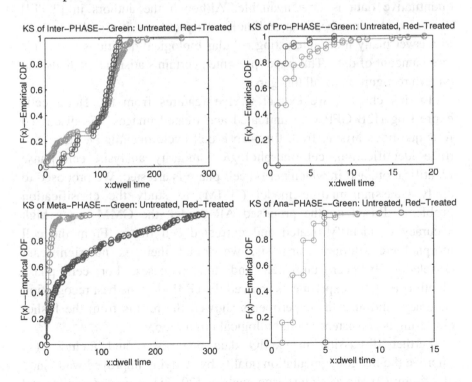

Figure 6: Kolmogorov–Smirnov test for interphase, prohase, metap5hase and anaphase. The curve shows the cumulative probability function vs. dwelling time.

is no significant difference between the cumulative probability functions of untreated and treated cell lines for interphase and prophase, but there is a significant difference between the untreated and treated cell lines for metaphase. That means many cells are arrested in metaphase. From the last sub figure of Fig. 6, we can see that lots of cells are dead in treated cell lines.

6. Conclusion

Time-lapse fluorescence microscopy imaging provides an important method to study the cell cycle process under different perturbations. Current methods, however, are rather limited in dealing with such time-lapse imaging datasets, whereas manual analysis is unreasonably time-consuming. With automated methods development, a large amount of quantitative data is now available. Although the authors in [2, 14] discussed certain subcellular and molecular problems, it is still open how to answer many more interesting cellular biological questions using this large amount of data. This paper presented certain statistical methods and pattern recognition to fill this gap.

In this chapter, we extract twelve features from the HeLa cells expressing H2B-GFP with untreated and treated images. We discussed four questions arising from time-lapse cell cycle quantitative data: cell trace identification, cell morphologic similarity analysis, cell phase identification, and in which period cell phase is arrested. We proposed to apply Gaussian mixture model (GMM) to study the classification problems. Based on the proposed AR model, the GMM has a high accuracy to identify treated and untreated cell traces. From the cell morphologic similarity analysis, we found there is no significant correlation between untreated and treated cases. For cell phase identification, the experiments showed the GMM has the best recognition accuracy, and also the experiments showed the results from the Fisher clustering is consistence with biological observation.

Further discussion in CellIQ data analysis should include: (1) improve the image segmentation quality by studying improved watershed algorithm; (2) enhance clustering analysis [29, 31] of treated and treated CellIQ data based on cell features and cell phase information; (3) study more cell features; and (4) improve the accuracy of cell phase recognition using more sophisticated classification technologies, such as context-based GMM and fuzzy support vector machine [18] and ensemble classifiers [17]. We did not discuss the feature reduction in this chapter. If there are more features, say about tens of features available, then we will investigate feature selection techniques to study cell phase identification [2, 14, 19, 32, 33].

Appendix A

First, the samples are partitioned into Q clusters using fuzzy C-means clustering, and then initial parameters are estimated using the standard vector quantization method [25]. We then estimate the parameters in (5) using the expectation-maximization (EM) algorithm [21] by iterating the following steps:

$$\gamma_k(\mathbf{x}) = \frac{w_k \mathrm{N}(\mathbf{x};\boldsymbol{\mu}_k,\Sigma_k)}{\sum_{i=1}^{K} w_i \mathrm{N}(\mathbf{x};\boldsymbol{\mu}_i,\Sigma_i)}, \qquad for\ all\ \mathbf{x} \in \mathbf{X}_m$$

$$\beta_k = \sum_{\mathbf{x} \in \mathbf{X}_m} \gamma_k(\mathbf{x})$$

$$w_k = \beta_k \bigg/ \sum_{i=1}^{K} \beta_i$$

$$\boldsymbol{\mu}_k = \sum_{\mathbf{x} \in \mathbf{X}_m} [\gamma_k(\mathbf{x})\mathbf{x}] \beta_k$$

$$\Sigma_k = \mathrm{diag}\left\{ \sum_{\mathbf{x} \in \mathbf{X}_m} \gamma_k(\mathbf{x})(\mathbf{x}-\boldsymbol{\mu}_k)(\mathbf{x}-\boldsymbol{\mu}_k)^T \right\} \bigg/ \beta_k$$

where $\mathbf{X}_m$ is the training sample set of class m. It has been shown that under some relatively general conditions, the iteration convergence is only guarantees, at least locally. Considering computation complexity, Σ_k is used as a diagonal matrix in our studied model, namely, $\Sigma_k = \mathrm{diag}\{\sigma_{k1},\sigma_{k2},...,\sigma_{kn}\}$. We have succussed in applying mixture model to gene microarray study [30] and other classification problems [34].

Appendix B

The Fisher clustering algorithm is a dynamic optimal algorithm to find the partition points without changing the order of the data. The algorithm is summarized as follows.

- Calculate $D(i, j)$ for all $1 \leq i \leq n$ and $1 \leq i \leq j \leq n$.
- Calculate the minimum of the objective clustering function as

$$e[P(i,2)] = \min_{2 \leq j \leq i}\{D(i, j-1) + D(j,i)\}, \quad 2 \leq i \leq n.$$

- Obviously it is the minimum of the objective clustering function when to partition the n time points into 2 clusters. When to partition the n time points into $k(k > 2)$ clusters, we then iteratively calculate the minimum of the objective clustering function as

$$e[P(i,k)] = \min_{l \leq j \leq i}\{e[P(j-1,l-1)] + D(j,i)\}, \quad 3 \leq l \leq k, 2 \leq i \leq n.$$

- Search the optimal partition points. When we partition the n time points into k clusters, if there exists a $j(k-1 \leq j \leq n)$ such that

$$e[P(n,k)] = e[P(j-1,k-1)] + D(j,n),$$

then time points $(j, j+1,...,n)$ is the first class, furthermore, we continue to partition the time points $(1,2,..., j-1)$ into k-1 clusters, if there exists j' such that

$$e[P(j-1,k-1)] = e[P(j'-1,k-2)] + D(j', j-1)],$$

then time points $(j', j'+1,..., j+1)$ is the second class. Then repeating the above steps, and finally we can get the all classes.

For more details about the discussion of this algorithm, we refer to [11, 19].

Acknowledgement

The authors would like to acknowledge the collaboration of Drs. Randy King and Susan Lyman, Department of Cell Biology and ICCB (Institute of Chemistry and Cell Biology), Harvard Medical School. They also provided the raw datasets for this study. This research was supported by the Center for Bioinformatics Program grant of Harvard Center of Neurodegeneration and Repair, Harvard Medical School, Boston, USA and a NLM R01 LM008696-01 to STCW.

References

[1] Bleau A. and Leon J.L., "Watershed-based segmentation and region merging," Computer Vision and Image Understanding, 2000; 77:317-370.

[2] Boland M.V. and Murphy R.F. "A Neural Network Classifier Capable of Recognizing the Patterns of all Major Subcellular Structures in Fluorescence Microscope Images of HeLa Cells," Bioinformatics, 2001; 17:1213-1223.

[3] Chai, B.B., Huang T., Zhuang X., Zhao Y., and Sklansky J., "Piecewise linear classifiers using binary tree structure and genetic algorithm," Pattern Recognition, 1996; 29:1905-1917.

[4] Chen S. and Haralick M. "Recursive erosion, dilation opening and closing transform," IEEE Transactions on image processing, 1995; 4:335-345.

[5] Chen X. Zhou X., and Wong S.T.C., "Automated segmentation, classification, and tracking cancer cell nuclei in time-lapse microscopy," accepted for publication in IEEE Trans on Biomedical Engineering.

[6] Davis L., ed. Handbook of Genetic Algorithm. Van Nostrand Reinhold, New York, 1991.

[7] Edwards, A.L. An Introduction to Linear Regression and Correlation. San Francisco, CA: W. H. Freeman, pp. 33-46, 1976.

[8] Endlich B., Radford I.R., Forrester H.B., and Dewey W.C. "Computerized video time-lapse microscopy studies of ionizing radiation-induced rapid-interphase and mitosis-related apoptosis in lymphoid cells," Radiat Res., 2000; 153:36-48.

[9] Goldberg D.E. Genetic Algorithms in Search, Optimization and Machine Learning. Addison-Wesley, Reading, Massachusetts, 1989.

[10] Gunst R.F. and Mason R.L. Regression analysis and its application: a data-oriented approach. New York: M. Dekker, 1980.

[11] Hantigan J.A., Clustering Algorithms, John Wiley & Wiley Sons, 1975.

[12] Haraguchi T., Ding D.Q., Yamamoto A., Kaneda T., Koujin T., Hiraoka Y., "Multiple-color fluorescence imaging of chromosomes and microtubules in living cells," Cell Struct Funct, 1999; 24:291-298

[13] Hiraoka Y and Haraguchi T, "Fluorescence imaging of mammalian living cells," Chromosome Res, 1996; 4:173-176.

[14] Huang K. and Murphy R.F. "Boosting accuracy of automated classification of fluorescence microscope images for location proteomics," BMC Bioinformatics, 2004; 5:78.

[15] Kanda T., Sullivan K. F. and Wahl G. M, "Histone-GFP fusion protein enables sensitive analysis of chromosome dynamics in living mammalian cells," Current Biology, 1998; 8:377-85.

[16] Karlsson C., Katich S., Hagting A., Hoffmann I., and Pines J. "Cdc25B and Cdc25C differ markedly in their properties as initiators of mitosis," Journal Cell Biology, 1999; 146:573-584.

[17] Kaufman L. and Fousseeuw P. J. Finding Groups in Data: An Introduction to Cluster Analysis. New York Wiley. (1990).

[18] Kecman V. Learning and Soft Computing. The MIT Press, Cambridge, MA, 2001.

[19] Kittler J. Feature selection algorithm, pattern recognition and signal processing. Germany: Sijthoff & Noordhoof, (1978) 41-60.

[20] Makhoul J., "Linear prediction: a tutorial review," Proceedings of IEEE, 1975; 63:561-580, 1975.

[21] G. Mclachlan and T. Krishnan, The EM Algorithm and Extensions, John Wiley & Sone, New York, 1997.

[22] Norberto M., Andres S., Carlos Ortiz S. Juan Jose V., Francisco P., and Jose Miguel G., "Applying watershed algorithms to the segmentation of clustered nuclei," Cytometry, 1997; 28:289-297.

[23] Otsu N. "A threshold selection method from gray level histogram," IEEE Transactions on System man Cybernetics, 1978; 8:62-66.

[24] Ripley B.D. Pattern Recognition and Neural Networks. Cambridge University Press. 1996.

[25] Rabiner L.R., "A tutorial on hidden Markov models and selected applications in speech recognition," Proc. of IEEE, 1989; 77:257-286.

[26] Specht D.F. "A generalized regression neural networks," IEEE Trans on Neural Networks, 1991; 2:568-576 .

[27] Valentini, G. and Masulli, F. "Ensembles of learning machines," in Neural Nets WIRN Vietri-02, Series Lecture Notes in Computer Sciences, Marinaro M. and Tagliaferri R., Eds.: Springer-Verlag, Heidelberg (Germany), 2002.

[28] Windoffer R. and Leube R.E. "Detection of cytokeratin dynamics by time-lapse fluorescence microscopy in living cells," Journal Cell Science, 1999; 112:4521-4534.

[29] Zhou X. and Chen Q., "Adative mixed differential clustering algorithm," Journal of Practice and Application of Mathematics, 1997; 27:312-319.

[30] Zhou X., Wang X., and Dougherty E.R., "Binarization of microarray data based on a mixture model," Molecular Cancer Therapeutics, 2003; 2:679-684.

[31] Zhou, X., Wang, X., Dougherty, E. R., Russ, D., and Su, E., "Gene clustering based on cluster-wide mutual information," Journal of Computational Biology, 2004; 11:147-161.

[32] Zhou X., Wang X., and Dougherty E.R., "Nonlinear probit gene classification using mutual-Information and wavelet," Journal of Biological Systems, 2004; 12:371-386.

[33] Zhou X., Wang X., and Dougherty E.R., "Sequence Monte Carlo estimation for nonlinear-probit gene classification based on bootstrap Bayesian gene selection," Journal of Franklin Institute, 2004; 341:137-156.

[34] Zhao Y., Zhou X., Palaninappan K., and Zhuang X., "Statistical modeling for improved land cover classification," Proc. of SPIE on Aerospace/Defense Sensing, Simulation, and Controls, Vol. 4741. Orlando FL, 2002.

CHAPTER 3

DIVERSITY AND ACCURACY OF DATA MINING ENSEMBLE

Wenjia Wang

School of Computing Sciences, University of East Anglia, Norwich, UK

Abstract

An ensemble can have a better overall performance than individual models that work alone if its member models are diverse enough from each other. However, a high level of diversity among the models does not come easily and the ensembles built with them may not produce any gain at all. This chapter firstly describes why diversity is essential and how the diversity can be measured. Then it analyses the relationships between the accuracy of an ensemble and the diversity among its member models and the individual's accuracy, and the number of models that are required to make an ensemble effective.

The chapter then move to explore possible strategies to enhance diversity, and two approaches, i.e. combining different modelling methodologies and decimating input features, are found to be effective in avoiding coincident failures. For demonstrations, two kinds of inductive techniques, i.e. neural networks and decision trees, are employed to implement different types of ensembles. Their performance and diversity are assessed and the results show that the heterogeneous ensembles have relatively high-level of diversity and thus are able to improve their performance. The described ensemble methods are applied to osteoporosis detection problem.

1. Introduction

It is generally accepted that for a non-trivial data-defined problem, single models that are generated from machine learning techniques may not be able to model the every aspect of the problem, and perform satisfactorily when work alone. Then ensemble approach in which multiple models are combined to work collectively has been explored by many researchers such as Hansen *et al.* (1990), Krogh *et al.* (1995), Partridge *et al.* (1997), Wang *et al.* (1998), Dietterich (2000), Kuncheva *et al.* (2003), Melville & Mooney (2004), Brazier & Wang (2004). The key idea behind is that a collection of multiple models combined together will have a better overall performance than that of the individual models. However, the studies carried out by Eckhardt *et al.* (1985), Littlewood & Miller (1989) and Partridge *et al.* (1996) have shown that it is not always the case simply because that the models even developed 'independently' of each other are likely to fail dependently. In other words, dependent models are likely to make the same mistakes and then an ensemble built with such models is unlikely to have an improved accuracy. To make an ensemble more accurate, its member models, apart from having a certain level of accuracy, must be diverse enough from each other to prevent making common failures simultaneously. Nevertheless, a high level of diversity does not come easily by just manipulating some modelling parameters when generating models as member candidates for building an ensemble and hoping it will be beneficial, without really understanding the relationships between ensemble's performance and diversity and accuracy of individual models.

This chapter attempts to address these problems and reveal these relationships. It starts with an illustration by means of a simple example of two ensembles to demonstrate the necessity of diversity. Then it moves on to describe how the diversity is measured. Section 5 focuses on analysing the relationships between ensembles and individual models. Section 6 describes the methods for building ensembles, two methods in particular: input decimation, and hybridisation of decision trees and neural nets. It follows with an application for a healthcare problem - osteoporosis classification. The chapter ends with discussion and conclusions.

2. Ensemble and Diversity

2.1. *Why need diversity?*

While significant attempts have been made to generate individual models as accurate as possible, and when their accuracy is usually still not good enough, attentions are paid to how to make the models diverse from each other to make an ensemble more accurate.

A simple example is presented below to illustrate why diversity among models is essential when they are used to build an ensemble. Assume that an ensemble is built with only three models for a classification problem and a simple majority-voting strategy is therefore employed to make the final decision of the ensemble. In addition, assume that each model has been trained to produce an accuracy of only two thirds on the data set of the problem, or they make errors on one third of the data. Then two Ensemble A and B are built with 3 models each for this classification problem, as shown in Fig. 3.1.

Ensemble A [Fig. 3.1(a)] is built with three identical models, i.e. classifiers in this case, which obviously have no diversity from each. Each model gives the right results for the same 2/3 of the data samples and makes the wrong decisions on the identical 1/3 remaining examples. Then with the simple majority voting on the outputs of the three classifiers, the ensemble will also produce 2/3 correct classification decisions on the data, an identical accuracy as that of the individual models. So, this ensemble makes no gain at all and should not be built in the first place.

On the other hand, Ensemble B [Fig. 3.1(b)] is also built with 3 models each with the same accuracy, i.e. 2/3, but each of them makes 1/3 errors different from each other. It is clear that for every test data instance, there are always two classifiers giving the right classification, and one giving the wrong decision. Then, by using the majority-voting strategy, the ensemble will have correct majority votes on every data instances and thus always yield the correct result, i.e. 100% accuracy, although each model working individually only has an accuracy of about 67.7%. Such an ensemble is considered to be ideal because its members have a maximum diversity.

W. Wang

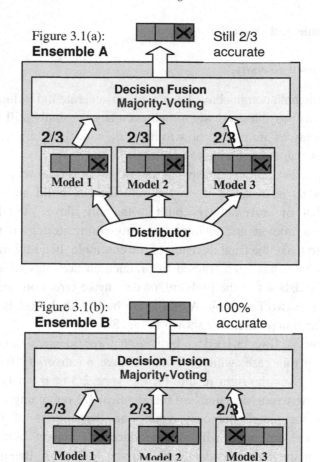

Figure 3.1: (a) Ensemble A built with 3 models which have the same 2/3/ accuracy but no diversity from each other. (b) Ensemble B is built with 3 models which still have 2/3 accuracy but make 1/3 errors different from each other.

This simple example of two ensembles, thought extreme, clearly illustrated the necessity of diversity among individual models when build an ensemble with less accurate models. Theoretical analysis and empirical studies [Eckhardt *et al.*, 1985; Littlewood & Miller, 1989; Hansen & Salamon, 1990; Partridge *et al.*, 1996; Wang *et al.*, 2001a] pointed out that (i) an ensemble will be beneficial iff its member models are diverse; (ii) when the models of an ensemble have a maximum diversity and an accuracy better than a random guess, the ensemble will achieve 100% accuracy.

However, generating and measuring the diversity is very challenging for complex problems. In particular, when the models are trained by machine learning algorithms, the models tend to be closely correlated thus no or very little diversity can be introduced. Then an ensemble built with those models will hardly produce any gain at all.

2.2. *Diversity measures*

Diversity may be measured in different ways. [Kuncheva *et al.*, 2003] summarised ten popularly used diversity definitions from pair-wised correlations to Entropy measures. This chapter examines the effectiveness of some commonly used diversity measures, which are presented in this section for convenience.

Notations:
   Data vector: $\mathbf{d}= [\mathbf{x}, y]$, Inputs: $\mathbf{x}=\{x_1, x_2, ...x_n\}$, Output: y
   Ω: the number of data instances in the data set
   Θ: model (classier or predictor)
   A: accuracy of a model or an ensemble
   D: diversity
   Φ: Ensemble
   M : the number of test data instances
   N: number of models in ensemble Φ.
   k_n: the number of examples that fail on exactly n models in Φ.
   p_n: probability that exactly n models fail on a randomly selected test data x

3. Probability Analysis

[Littlewood & Miller, 1989] defined a number of diversity measures in terms of probability of simultaneous failure, such as, $P(\Theta)$ — the probability that a randomly selected model Θ, from Φ fails on a randomly selected input $\mathbf{x}$; $P(\Theta^2)$ — the probability of two models selected at random from Φ will both fail on a random input, and variance $\mathrm{Var}(\Theta) = P(\Theta^2) - P(\Theta)^2$. These definitions were derived on assumption of infinite sets of models and examples, and intend to measure independence of failures. In reality, both the numbers of models and input instances are likely to be small in some cases.

4. Coincident Failure Diversity

Partridge *et al.* (1997) defined some other measures based on probability analysis.

- The probability p_n that exactly n models fail on a randomly selected test data x is defined as:

$$p_n = k_n \Big/ M, \quad n = 1, \ldots, N \tag{3.1}$$

- The probability that a randomly selected model from A fails on a randomly selected input:

$$p(1) = \sum_{n-1}^{N} \frac{n}{N} p_n, \tag{3.2}$$

Similarly,
- The probability that two models selected from Φ at random (without replacement) both fail on a randomly selected input:

$$p(2) = \sum_{n-1}^{N} \frac{n(n-1)}{N^2} p_n, \tag{3.3}$$

In general, the probability that r randomly chosen classifiers fail on a randomly chosen input can be formulated as:

$$p(r) = P(r \text{ randomly chosen models fail})$$
$$= \sum_{n=1}^{N} \frac{n}{N} * \frac{(n-1)}{(N-1)} * \ldots \frac{(n-r+1)}{(N-r+1)} * p_n \qquad (3.4)$$

In addition, the probabilities, e.g. $p(1 \text{ out of 2 is correct})$ or simply $p(1/2)$, that 1 out of 2 randomly selected classifiers produces correct answer, $p(2 \text{ out of 3 is correct}) = p(2/3)$, etc, have also been used for measuring the stability of an ensemble system.

The essence of developing multiple model systems is to reduce chance that members in a system make mistakes coincidentally. Therefore, a diversity that measures coincident failures of N members on the same input can be estimated by the following equation.

$$CFD \equiv \begin{cases} \dfrac{1}{1-p_0} \sum_{n=1}^{N} \dfrac{N-n}{N-1} p_n, & if \ p_0 < 1 \\ 0, & if \ p_0 = 1 \end{cases} \qquad (3.5)$$

$CFD \in [0, 1]$. When $CFD = 0$, it indicates either all failures are the same for all models—hence no diversity; or there is no test failure at all, i.e. all models are perfect and identical-hence no diversity (there is no room for diversity if all models are perfect, and thus on need for building ensembles). $CFD=1$ when all test failures are unique to one model, i.e. p_1 $=1$, in this case the ensemble of those models will be ideal, i.e. always produce correct answers, but in reality it will be extremely difficult to construct an ideal ensemble with a maximum CFD diversity.

Another measure was derived when a simple majority-voting decision strategy is applied, from the probability that the majority of the models $(N+1)/2$ in Φ produce the correct answer for a randomly chosen input [Wang *et al.*, 1998]:

$$p(maj) = 1 - \sum_{n=(N+1)/2}^{N} p_n = \sum_{n=0}^{(n-1)/2} p_n \qquad (3.6)$$

Table 3.1. Diversity and performance assessments of two Ensembles of the illustrative example and their member models.

	Ensemble A		Ensemble B	
(a) Performance of 3 individual models				
Model	Success	Fail	Success	Fail
1	66.7%	33.3%	66.7%	33.3%
2	66.7%	33.3%	66.7%	33.3%
3	66.7%	33.3%	66.7%	33.3%
(b) Coincident failures of the models in A and B				
n/N	k_n	p_n	k_n	p_n
0/3	2	0.667	0	0.0
1/3	0	0	3	1.0
2/3	0	0	0	0.0
3/3	1	0.333	0	0.0
(c) Ensemble performance: diversity, accuracy				
p(1)	0.333		0.333	
p(2)	0.333		0.000	
p(1/2)	0.667		1.000	
p(2/3)	0.667		1.000	
CFD	0.000		1.000	
p(maj)	0.667		1.000	
Voting	**66.7%**		**100%**	

These measures are applied to the two Ensembles of the early illustration example. The assessment results, in terms of diversity and performance, for the individual models and the 2 ensembles are listed in Table 3.1.

Part (a) of Table 3.1 shows the performance of the models that work individually. They all have the same success rate (accuracy) - a mere 66.7%.

Part (b) shows the coincident failures of the models in Ensemble A and B respectively. Each row tells that k_n data examples failed exactly on n models, and its corresponding probability p_n. e.g. row 0/3, for

Ensemble A, two test data examples failed exactly on 0 classifier of 3 classifiers, and the probability p_0 is 0.667; or in other words, all 3 classifiers produced correct results for the 2 test data examples.

Part (c) summarises the ensemble performance. It can be seen that Ensemble A has no diversity (CFD=0) at all thus no gain achieved. But for Ensemble B, CFD=1, a maximum diversity, so it produces 100% accuracy by majority-voting.

5. Ensemble Accuracy

Having just illustrated the importance of diversity in influencing the accuracy of an ensemble, it is worth noting that some other factors should also be considered when computing the accuracy of ensemble Φ, which are, the accuracy A of individual models, N - the number of the models in Φ, and S – the ensemble decision fusion strategy. This can be represented by

$$A(\Phi) = f (A(\Theta, ..), D, N, S) \qquad (3.7)$$

This section will present some theoretical as well as empirical analysis results for some special cases.

5.1. *Relationship between random guess and accuracy of lower bound single models*

As mentioned before, the models should have accuracy better than a random guess, thus the ensemble built with them is able to improve accuracy, provided they are diverse enough. This random guess can then be used as the accuracy lower bound to for evaluating the models, which can be estimated, for a classification problem, by:

$$Alb(\Theta) = \lim_{N \to \infty} \frac{N+1}{CN} = \frac{1}{C} \qquad (3.8)$$

Where, C is the number of output classes. The relationship between C and the accuracy lower bound *Alb* is shown in Fig. 3.2, when C = 4, Alb=0.25.

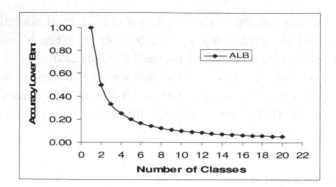

Figure 3.2: The lower bound of accuracy of individual models for different number of target classes.

When this bound is known, the models/classifiers can assessed if they are suitable in term of accuracy as the candidates for building an ensemble.

5.2. *Relationship between accuracy A and the number of models N*

Another relationship that is very useful for guiding the construction of ensembles and should be investigated, is that, given a pool of model candidates and their diversity and accuracy, what the minimum number of the models are needed to build a best ensemble. As there are many variations involved with diversity and accuracy, it is difficult to give a complete answer. However, for some special cases, for instance, assuming that the diversity among the models has been estimated and kept constant, and the lower bound of the accuracy Alb of individual models in the pool is known, then we want to know the minimum number of the models needed for building an ideal ensemble, which can be represented by $N_{min}=f(Alb\{\Theta\} \mid D)$. The results of our empirical studies are shown in Fig. 3.3.

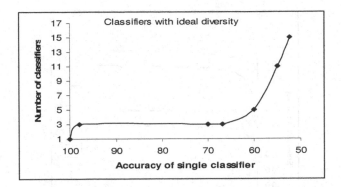

Figure 3.3: The relationships between the accuracy lower bound of individual classifiers and the number of the classifiers needed for building an ideal ensemble.

5.3. When model's accuracy < 50%

In literature, it is generally accepted that the accuracy of individual models that are used for building an ensemble should have an accuracy better than a random guess 1/C, e.g. 1/2 for dichotomous problems, then the ensemble can be beneficial [Hansen & Salamon, 1990]. This study proves that even the accuracy of individual models is less than 1/C, the ensemble can also improve its overall accuracy, provided that the models (classifiers) are diverse enough in the right places. Assume that the models of an ensemble Φ have ideal diversity, then, the highest accuracy of the ensemble can be estimated by:

$$\forall \ A(\Theta_i) \le 50\%:$$

$$A_{\max}(\Phi) = \lim_{N \to \infty} \frac{2N\,A(\Theta)}{N+1} = 2A(\Theta) \qquad (3.9)$$

It simply indicates that the maximum accuracy of an ensemble Φ will eventually reach to as much as double of the accuracy lower bound of the models in Φ iff they have ideal diversity. Figure 3.4 depicts the relationships between the accuracy of ensembles and the minimum number of the models required to build the ensembles, given different accuracy lower bounds (from 50 down to 25%) of the models and ideal diversity among them.

58 W. Wang

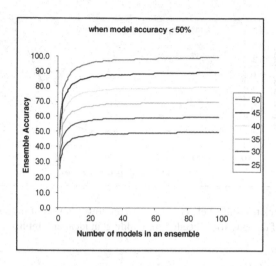

Figure 3.4: The relationships between ensemble accuracy and the minimum number of the models that are needed to build the ensemble, where the accuracy lower bound of the models is less than a random guess (i.e. 50% for binary classification).

6. Construction of Effective Ensembles

Methods have been explored in attempting to build better ensembles by trying either to generate more accurate models or to create more diverse models, or both ideally. Bagging [Breiman, 1996] and Boosting [Schapire *et al.*, 1997] or Adaboots [Freund & Schapire, 1996] are two popularly used techniques to manipulate the data set by adding more weight to so called 'hard' data subsets to force the models to learn the aspects represented by these weighted training data subsets. A study carried by Optiz & Maclin (1999) compared their performance and found a Bagging ensemble generally produces a classifier that is more accurate than a standard classifier, the Boosting, however, produced widely varying results largely due to its high sensitivity to the noise level in the data.

In addition, the decision fusion strategies also play important role in determining the performance of an ensemble. *Averaging* and simple or weighted *voting*, are the tow commonly used ones, pending the type of machine learning algorithms employed. For one, such as neural networks, outputs continuous value, *averaging* seems naturally suitable but the diversity, if not carefully handled, may have some adverse effects on the final averaged result. *Voting* strategy is best suitable for the modelling algorithms with categorical outputs, such as decision trees for classification problems. However, the continuous outputs can be easily converted to the discrete decisions, then *voting* can be applied.

6.1. *Strategies for increasing diversity*

Many strategies and tactics have been explored in attempt to increase diversity when developing models, which include altering learning parameters, initial conditions, model structure, learning stop conditions, manipulating data set such as Bagging [Breiman, 1996]. However, the studies [Partridge *et al.*, 1996; Wang *et al.*, 1998] show that, in the case of combining the models of same machine learning methodology, such as neural networks, or decision tress, it is not easy to achieve high level diversity between the trained neural nets by just manipulating initial conditions, architecture and learning parameters due to the natural similarity of supervised training algorithms. The gain on diversity by these strategies is very limited. In terms of diversity generated, the study gave a ranking of them in the following order with the most diverse strategy first: type of neural nets > training sets > architecture of nets >= initial conditions. The studies by [Littlewood & Miller, 1989] indicated that models implemented by different methodologies may produce a higher level diversity than other variations. Neural networks and decision trees are different learning methods and therefore a combination of them is potential to achieve diverse and better systems. In addition, employing the technique of input decimation and data partitions could provide further improvement on diversity, which can be justified by the way neural nets are developed. Neural nets are trained to learn from the data provided so they are data-dependent. Different data set may represent different dominant knowledge of the problem. This study uses manipulation of the number of the dimensions by choosing different input features.

- ### Input decimation based on estimate of feature salience

Above results indicate that using different data subsets could generate more diversity than manipulating the other parameters. Tuner & Oza (1999) tried to use different subsets of input features in various sizes to train nets and reported the improvement on the performance of multiple model systems they subsequently built. However, in the latter study they selected the features according to correlation coefficient of features to the corresponding output class. Wang *et al.* (1998, 2001) found that selecting features with such a method usually produces poor results except for some simple, linear problems. We developed some other techniques for identifying the salience of input features and the comparative study [Wang *et al.*, 2000] indicates that our techniques performed better for complicated, noise real-world problems.

- ### Estimating input feature salience

A number of techniques we investigated for identifying salience of input features are developed based primarily on neural network technology. Two of them are our own proposed methods, i.e. neural net clamping and the decision tree heuristic algorithm [Wang *et al.*, 1997 1998 2001, 2002]. The other two methods are input-impact analysis [Tchaban *et al.*, 1998] and the linear correlation analysis, which is used as a baseline.

- ### Input decimation

The features are ranked in accordance with their value of salience, i.e. the impact ratio in the clamping results or significant score with decision tree heuristic. The larger are the value the more salient the features. Then different numbers of salient input features can be selected from the top of ranks to form subsets of features for training models.

6.2. Ensembles of neural networks

The technique of a multiple model system of neural nets has been widely used in various applications. However, a multiple model system does not always produce better performance because the members in such a system, i.e. the trained neural nets are highly correlated and tend to make the same mistakes simultaneously. So much attention and effort have been put into finding methods that could create diversity between nets.

This study focused on investigating constructing multiple model systems with the multi-layer perceptrons neural nets trained by using different subsets of input features (even initial weights and number of hidden units are also varied). Only multi-layer perceptrons are utilised here primarily because they are the most popularly used type of neural nets in the constructions of multiple model systems. The procedure we used for building a multiple net system is below. For each input decimated data set:

   (i) Partitioning the data set with the procedure described in the earlier section.

   (ii) Training and validating neural nets with the training and validation data sets respectively. (Three nets designed with 3 different number of hidden nodes were trained with each training set using different initial conditions. Thus a cube of 27 nets in total were produced and placed in a pool as the candidates of models.)

   (iii) Constructing multiple net systems by randomly selecting the nets from the candidate pool.

   (iv) Estimating the diversity and assessing the performance with the test data set.

6.3. *Ensembles of decision trees*

A decision tree is a representation of a decision generating procedure induced from the contents of a given data set. Induction of a decision tree is a symbolic, supervised learning process using information theory, which is methodologically different from the learning of neural networks. We use the C5.0 decision tree induction software package (as a black box) -- the lasted development from ID3 [Quinlan, 1986], to generate the member candidates of trees. Moreover, a heuristic algorithm is developed based on the trees induced to determine the salience of input features for input decimation.

The procedure for building a multiple tree system is the same as the one for building multiple net systems except that the candidates are the decision trees induced by varying pruning levels.

6.4. *Hybrid ensembles*

With two different candidates of models, i.e. neural nets and decision trees, available, it is now possible to combine them to construct hybrid multiple model systems. A specific mechanism was designed to build a combined ensemble, in which the number of models from each type is controlled by a given ratio. For an N model system, m of N models must come from a designated candidate type pool and the remaining $(N-m)$ from another. For instance, if m is the number of trees in an ensemble, altering m from 0 to N, we can obtain a set of hybrid systems that composed of different numbers of trees (from 0 to N) and a complement number of nets (from N to 0). In fact, when m=0, the hybrid systems become the multiple model systems purely composed of neural nets. On the other hand, when $m=N$, the systems are the pure multiple tree systems. In this way, the previous two types of multiple model systems are just two special cases of hybrid systems.

7. An Application: Osteoporosis Classification Problem

We have applied our techniques to a number of real-world problems after tested them on some artificial problems. Here we present the results of one of those real-world problems, i.e. osteoporosis disease classification. In addition, a commonly used technique in healthcare field, i.e. Logistic Regression is also used to provide a baseline for comparisons.

- **Assessment of classification accuracy**

The classification accuracy is usually measured by the generalisation accuracy. However, this measure has a serious drawback, i.e. its value does not really indicate the true performance of a classifier when a test data set is unbalanced. For instance,

For dichotomy classification problems the Receiver Operation Characteristic (ROC) curve is an effective measure of performance. It can avoid the above drawback by showing the sensitivity and specificity of classification for two classes over a complete domain of decision thresholds. The sensitivity and (1-specificity) at a specific threshold value or/and the area under the curve are usually quoted as indicators of performance.

7.1. *Osteoporosis problem*

Osteoporosis is a disease that causes bones to become porous and to break easily. Identification of the most salient risk factors for the disease and the use of these risk factors for predicting the disease development will be very helpful for medical profession. A data set of over 700 cases was collected from regional hospitals when they were referred to osteoporosis clinics for tests by screening equipment, i.e. Quantitative Ultrasound (QUS).

The data contains 31 risk factors identified initially by the medical field experts as relevant to the disease. The diagnosis decision *osteoporotic* or *non-osteoporotic* is made based on the T-test score of the screening test. It can then be treated as a classification task and therefore taken as one of applications of our techniques.

7.2. *Results from the ensembles of neural nets*

Table 3.2 shows evaluation results for an ensemble of 9 neural nets, in the same format as that of Table 3.1.

Part (a) lists the performance of 9 single classifiers, i.e. generalisation, sensitivity and specificity which are obtained when set the threshold 0.5. The average generalisation rate is 0.74.

Part (b) shows the coincident failures of the 9 classifiers in this ensemble. Each row gives that k_n patterns failed exactly n classifiers, e.g. in row 3, 16 test patterns failed on exactly 2 classifiers of 9, and the probability p_2 is 0.08.

Table 3.2. Performance of Ensembles.

(a) Individual performance

classifier	Gen.	sensitivity	specificity
1	0.765	0.516	0.853
2	0.740	0.438	0.882
3	0.760	0.500	0.875
4	0.740	0.438	0.882
5	0.765	0.516	0.882
6	0.760	0.516	0.875
7	0.705	0.531	0.787
8	0.725	0.422	0.868
9	0.700	0.516	0.787

(b) Classifier coincident failures

n	k_n	p_n
0/9	121	0.605
1/9	4	0.020
2/9	16	0.080
3/9	2	0.010
4/9	11	0.055
5/9	4	0.020
6/9	6	0.030
7/9	3	0.015
8/9	4	0.020
9/9	29	0.145

(c) Ensemble performance

$P(\Theta)= 0.2533$,	$P(\Theta^2)= 0.2056$
$Var(\Theta)= 0.1413$,	**CFD= 0.4035**
p(2 different both fail)	$= 0.1996$
P(1 out of 2 correct)	$= 0.8050$
P(1 out of 3 correct)	$= 0.8250$
P(2 out of 3 correct)	$= 0.7549$
P(1 out of 9 correct)	$= 0.8550$
G(averaging)	**= 0.7400**
G(voting)	**= 0.7700**

Part (c) summarises the ensemble performance defined in the earlier sections. It can be seen that the generalisation, is improved by 3% when *majority-voting* strategy, *G(voting)*, is employed, compared with that of the *averaging* strategy, *G(averaging)*. The magnitude of improvement is not significant because the Coincident Failure Diversity (CFD) among its 9 members is not high, only around 0.40.

Table 3.3. Evaluations of the ensembles of neural nets.

features selected	G(mean)		G(voting)		CFD	
	Average	s.d.	Average	s.d.	average	s.d.
5	0.715	0.003	0.725	0.003	0.223	0.070
10	0.732	0.005	0.777	0.006	0.465	0.036
15	0.734	0.004	0.757	0.008	0.455	0.010
20	0.700	0.003	0.712	0.003	0.345	0.074
mix-all	0.721	0.008	0.768	0.009	0.476	0.053

Table 3.3 summarises the evaluation results of multiple classifier systems that were built by respectively picking the classifiers from 4 pools of the nets trained with the top 5, 10, 15 and 20 salient features selected from original 31 features. The mixed Ensembles were also built by choosing 9 nets randomly from all 4 pools.

It should be noted that all numbers quoted in the Table are the average value over 27 ensembles, and their standard deviations (s.d.). *G(mean)* is the mean value of individual classifiers in the ensemble.

The results can be concluded as follows:

- The diversity CFD in the intra-category ensembles is considerably small (0.18 < CFD <0.35), which is already expected.
- The inter-category ensembles have larger CFD, which leads to some extent of improvement compared to the mean of the ensembles with *majority-voting* strategy applied.
- The generalisation obtained by the *majority-voting* strategy is only marginally better than that of ensembles with the *averaging* strategy, which suggest that these decision strategies behaved similar in this circumstance.
- The mixed ensembles do increase the diversities (0.40<CFD<0.51) but the generalisations of majority-voting are not increased proportionally because of lower *mean* generalisation over all the individual nets.
- The ensembles constructed with the nets from some specific groups (e.g. nets trained with 10-salient-feature data set) produce the best performance.

Table 3.4. Performance of the tree's ensembles.

pruning level(%)	worst tree	best tree	G(mean)	G(voting)	CFD
10	0.680	0.755	0.726	0.750	0.490
25	0.650	0.745	0.709	0.740	0.592
50	0.655	0.735	0.692	0.740	0.629

7.3. Results from ensembles of the decision trees

Table 3.4 summarises the results of evaluations on the ensembles that were only composed of the decision trees induced with all 31 input features (i.e. no input decimation) in the data by setting the pruning confidence level to 10%, 25% (the default level) and 50%.

It is obvious that the diversity CFD increases as the pruning level rises and the difference between the generalisations obtained by the *majority-voting* and *mean* also increases. The generalisations by *majority-voting* are about 2.4-4.8% higher than the mean value of members in those systems. The systems yield generalisation almost as good as that of the best individual classifiers in the systems even though the overall generalisation still remains almost the same.

Table 3.5 summarises the evaluation results for the multi-classifier systems that were constructed with the trees induced by using the selected features. The results indicate (i) the multi-classifier systems built with the trees trained by the 10 selected features out-performed the others; (ii) All the tree's ensembles have relatively higher CFD. The mixed ensembles yield the highest CFD (0.614). In general,

- Ensemble classifier systems of the decision trees have higher CFD diversity (than those of the neural nets), but
- The generalisation performance is no better (or worse) than the ensembles of the nets.
- Trees are much quicker to develop.

Table 3.5. Tree ensembles with input decimation.

features chosen	worst tree	best tree	G(mean)	G(voting)	CFD
10	0.710	0.785	0.748	0.780	0.574
15	0.685	0.750	0.720	0.750	0.557
20	0.660	0.750	0.715	0.745	0.592
mix-all	0.660	0.765	0.723	0.771	0.614

7.4. *Results of hybrid ensembles*

With *majority-voting* strategy applied, the ensembles built purely with the neural nets (i.e. nets = 9, trees = 0) have relatively lower diversity (CFD = 0.445 on average) but a little bit higher generalisation, *G(voting)* = 0.783, and the highest *G(average)* because of higher individual performance of the trained nets. On the other end, the ensembles built purely with the decision trees have higher diversity (CFD = 0.543 on average) but slightly lower generalisation *G(voting)* = 0.781, and much lower *G(average)*.

By introducing a certain number of the diverse trees to the nets-dominated ensemble, CFD is increased, which means that nets and decision trees have learned different knowledge embedded in the data set and that they are diverse. Hence, the generalisation with the *majority-voting* strategy is then improved (nearly 3% when number of the decision trees is just below the majority number i.e. 4). Nevertheless, when the number of the trees in the ensemble is further increased, the *voting*-performance of the ensembles on average deteriorates because the diversity decreases, but magnitude of reduction on the performance is very small. The middle line in the figure shows the average generalisation of the hybrid systems achieved by using *averaging* decision strategy. It shows that as the number of the trees in the ensembles increases, the *averaging*-performance of the ensembles deteriorates because of lower performance of the decision trees.

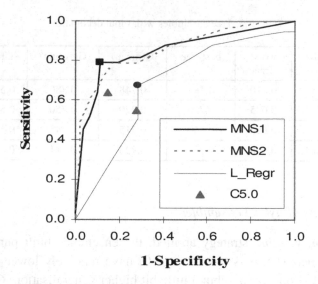

Figure 3.5: The ROC curves of two ensembles (MNS1 and MNS2) and the Regression method (L_Regr), and plots of the best individual decision tress. It can be seen that both ensembles outperformed the other methods.

Figure 3.5 shows the ROC plots for evaluating the performance of the ensembles, the individual classifiers, and the other techniques used for comparison purpose. It shows that ensembles produced some gains compared with the individuals but not really much. Nevertheless, they both well outperformed the linear regression method. The best cutting point (indicated by the solid square point on Fig. 3.5) on the curves gives that 80% correction classification for abnormal cases while has 10% misclassification for normal cases.

8. Discussion and Conclusions

This chapter investigates the relationships between the diversity and ensemble performance, gave the guidelines for choosing the individual models based on their accuracy and diversity measures for building good ensembles. The study demonstrates that an ensemble will be perfect iff its member models have an ideal diversity, but this requirement is too difficult to meet in reality for complex problems. The study also proved that even the models that have accuracy lower than a random guess it is still possible to build a beneficial ensemble that can, in principle, produce as much as twice of the lower bound accuracy of the individual models, provided again the models must be sufficiently diverse.

Furthermore, the hybrid ensembles of the trained neural nets and the automatically induced decision trees have been built for osteoporosis classification problem. The evaluation results presented in this chapter have shown that the classifiers trained with the data sets after input feature decimation are more diverse and performed better than those trained without feature selection. Higher performance achieved by the neural nets (or trees) trained with less salient features means that input decimation by ranking salient features was very successful not only in reducing the dimensions of the data set but also in improving the accuracy of classification. With a higher diversity the system is more reliable and produces consistent performance when tested with different data sets. However, the diversity between the trained neural nets is still not high enough to improve the performance significantly. The trees we induced appeared more diverse but have lower individual generalisation.

The ensembles constructed with the classifiers selected only from the same candidate pool (either trained neural nets or induced decision trees) can improve generalisation accuracy to a relatively small extent due to lower diversity within members of the pool when majority-voting strategy is applied. Combining neural nets and decision trees does create further diversity and consequently improve the accuracy of classification under the condition that the system is constructed with a majority of good nets and a minority of diverse-trees.

Further work will focus on theoretical analysis to prove the findings from the empirical studies presented in this chapter.

Acknowledgements

This research is currently funded by the EPSRC, UK, under grant numbers GR/R85914/01 and GR/R86041/01.

References

Brazier, K. & Wang, W. (2004): Implicit fitness sharing speciation and emergent diversity in tree classifier ensembles. Proc. of 5th Int. Conference on Intelligent Data Engineering and Automated Learning. pp333-338. Exeter, UK. August 2004.

Breiman, L. (1996): Bagging prediction. Machine Learning, Vol. 24, pp123-140.

Dietterich, T.G.(2000): Ensemble methods in machine learning. In Josef Kittler and Fabio Roli, editors, *Multiple Classifier Systems*, volume 1857 of *Lecture Notes in Computer Science*. Springer-Verlag, 2000.

Eckhardt, D. *et al.* (1985): A theoretical basis for the analysis of multiversion software subject to coincident errors. *IEEE Trans. Software Eng*. SE-11, pp1511-1517.

Esposito, R & Saitta, L.(2003): Monte Carlo Theory as an Explanation of Bagging and Boosting. Proceedings of the 18th Int. Joint conference on Artificial Intelligence, Morgan Kaufmann, pp499-504.

Freund, Y. & Schapire R.E. (1996): Experiments with a new boosting algorithm, in L. Saitta, ed., Machine Learning: Proceedings of the 13th national conference, Morgan Kaufmann. pp148-156.

Hansen, L. *et al.* (1990): Neural network ensembles. IEEE Trans. *Patterns Analysis and Machine Intelligence*, vol. 12 pp 993-1001.

Kuncheva, L.I.(2003): Measures of diversity in classifier ensembles and their relationships with the ensemble accuracy.. Kluwer Academic Publisher, Netherlands, 2003.

Kohavi, R & Wolpert, D.H (1996): Bias plus variance decomposition for zero-one loss functions. In Proc. ICML-96 Bari. Italy. pp275-283.

Krogh, K. & Vedelsby, J. (1995): Neural network ensembles, cross-validation and active learning. In G. Tesauro, D.S Touretzky, and T.K. Leen, eds, *Advances in Neural Information Processing System*. vol.7. pp231-238, MIT press.

Littlewood, B. & Miller, D. (1989): Conceptual modelling of coincident failures in multiversion software. *IEEE Trans. Software Eng*. vol. 15, no. 12, pp1596-1614.

Optiz, D. & Maclin, R. (1999): Popular Ensemble Methods: An Empirical Study. Journal of Artificial Intelligence Research, 11 1999, pp. 169-198.

Partridge, D. *et al.* (1996): Engineering multiversion neural-net systems. *Neural Computation*, vol. 8, pp869-893.

Partridge, D. & Krzanowski, W. (1997): Software diversity: practical statistics for its measurement and exploitation. *Information and Software Technology*, vol. 39, pp707-717.

Quinlan, J. R. (1986): Induction of decision trees. *Machine Learning*, 1, pp81-106.

Schapire *et al.* (1997): Boosting the margin: A new explanation for the effectiveness of voting methods. In Fisher, D.(Ed) Machine Learning: Proceedings of 14th Int. Conference, Morgan Kaufmann.

Tchaban, T., *et al.* (1998): Establishing impacts of the inputs in a feedforward network, *Neural Computing & Applications*, 7, pp309-317.

Tuner, K. & Oza, N. (1999): Decimated input ensembles for improved generalisation. Proceedings of IJCNN1999, Washington DC, October, 1999.

Wang, W. & Partridge, D. (1998): Multiversion neural network systems. Proceedings of neural networks and their applications, NEURAP'98, Marseilles, France, 1998, p351-357.

Wang, W., *et al.* (1998): Ranking pattern recognition features for neural networks. in S. Singh, ed. *Advances in Patterns Recognition*, Springer, pp232-241.

Wang, W., Jones, P. and Partridge, D.(2000): Assessing the impact of input features in a feedforward network. Neural Computing and Applications, Vol9, pp101-112, 2000.

Wang, W. Jones, P. and Partridge, D (2001): A comparative study of feature-salience ranking techniques, Neural Computation. MIT Press, Vol.13 (7), pp1603-1623, 2001.

Wang, W. & Partridge, D. and Etherington J. (2001a): Hybrid Ensembles and Coincident-Failure Diversity, in Proceedings of IJCNN'01, IEEE-INNS-ENNS. pp2376-2381. ISBN: 0-7803-7046-5. July 2001, Washington, USA.

Wang, W. (2002): A decision tree based heuristic method for identifying salience of risk factors. IDEAL'02 (IEEE/EPSRC). Manchester UK. Sept. 2002.

CHAPTER 4

INTEGRATED CLUSTERING FOR MICROARRAY DATA

Gabriela Moise and Jorg Sander

University of Alberta

Abstract

Various clustering algorithms have been applied to gene expression data in an attempt to reveal functionally related genes based on the assumption that genes with similar patterns of expression have similar biological function. Yet, there is no general consensus on which clustering algorithm performs best on gene expression data. In this chapter, we approach gene expression data clustering from a different perspective. Our approach falls under the umbrella of the much needed integration of the existing biological data sources. We take advantage on the wealth of microarray data being produced by integrating the results of multiple clusterings of the same data set under similar experimental conditions. Our experimental evaluation using the Gene Ontology (GO) shows that the "integrated" clustering result is biologically more meaningful than any of the clusterings being integrated.

1. Introduction

The rapid advance of various sequencing projects at genomic scale has triggered the development of DNA microarray technology that allows the study of the expression levels of thousands of genes in an organism simultaneously in a fast and efficient way. Analysis of the data produced by the microarray technology may reveal novel insights into gene functional annotation, tissue classification or genetic network reconstruction.

The raw microarray data is translated into gene expression matrices where typically the rows represent the genes; the columns represent the samples (e.g., tissues, experimental conditions, developmental stages), and an entry represents a measurement of the expression of a given gene in a given sample. One difficulty in analyzing gene expression data arises from the high dimensionality of the feature space that often exceeds the dimensionality of the sample space by a factor of 1000 or more. In addition, gene expression data is inherently noisy, due to biological and experimental variability. Thus efficient computational tools and visualization techniques are needed in order to mine knowledge out of these high dimensional data sets.

During the past few years, researchers have tried a large number of both unsupervised and supervised methods for the analysis of gene expression data. Clustering (i.e., grouping objects, such as genes or samples, based on their similarity in expression patterns) is arguably the most widely used unsupervised technique for gene expression data analysis. Different clustering algorithms have been applied to gene expression data, such as hierarchical clustering [6], k-Means [8], self-organizing maps [7]. In addition, there are several available software tools that can be readily used to analyze a given microarray data set through the means of clustering [23–25]. Major supervised analysis approaches include k-Nearest Neighbor (kNN) [10], Support Vector Machine (SVM) [9], and decision trees [11].

Clustering is used as a first step in the process of exploratory gene expression data analysis, which is essential when one deals with large amount of data on which there is little prior knowledge. Clustering gene expression patterns leverages the "guilt by association" assumption, i.e., genes with similar patterns of expression are likely to have similar biological function. Thus, the resulted clusters suggest potential functional relationships between genes that require further validation.

A well-known fact is that different clustering algorithms produce different clustering solutions on the same data, and even the same clustering algorithm may produce different clustering solutions on a given data set, due to different parameter settings or some randomness embedded within the algorithm. Multiple factors influence the choice of the "best" clustering algorithm for a given data set, e.g., data distribution, data dimensionality, or the impact of preprocessing techniques. Although significant effort has been spent on developing techniques for evaluating the quality of a clustering result [20], there is no general agreement on which clustering technique is most appropriate for gene expression data.

This paper takes a different approach with respect to clustering of gene expression data. There are several microarray experiments measuring the expression of a given organism's genes under similar experimental conditions. We attempt to enrich the knowledge concerning that organism's genes by integrating the knowledge obtained by clustering each individual microarray experiment.

For instance, consider yeast (*Saccharomyces cerevisiae*), which, having its genome fully sequenced, can be routinely measured in microarray experiments by virtually any lab. Several research labs may be interested in studying the yeast cell cycle, and therefore each lab would conduct its own microarray experiment for measuring the expression patterns of the yeast's genes during the cell cycle. The genes being measured in the microarray experiments are the same, but the experimental conditions, although they all capture the cell cycle, might slightly differ due to different strains of yeast that the labs are using, different synchronization procedures or different time points for recording the measurements. Each research lab applies its favorite clustering tool (or tools) and the most satisfying results are reported in the form of a clustering result, i.e., a partitioning of the set of genes into a set of disjoint, exhaustive and non-empty clusters. Each partitioning represents knowledge regarding the yeast's genes regulation during the cell cycle that has been extracted based on the microarray experiment. We are interested in investigating whether integrating the knowledge encoded by each partitioning can produce a biologically more relevant partitioning of the yeast's genes.

Note that yeast is not the only organism that has been studied in multiple microarray experiments. There are multiple microarray data sets for other organisms, such as mouse, C.elegans [28], or for different types of cancer in humans [29].

The approach described in this chapter falls within the more general field of biological data sources integration. Also related to the present work is the concept of cluster ensemble or consensus clustering. The cluster ensemble problem can be informally defined as the problem of improving the clustering performance by first generating multiple partitionings of a given data set, and then combining them in order to form a final (presumably superior) clustering solution. We discuss relevant related work in Section 2.

Having as a common denominator the integration of multiple partitionings, our approach differs from the cluster ensemble formalism in that we integrate partitionings derived from multiple data sets through various clustering algorithms, as opposed to integrating partitionings derived on the same data set, which are typically produced by different runs of the same clustering algorithm (e.g., K-Means). Note that - in contrast to the cluster ensemble framework - our approach integrates the *results* of multiple clustering algorithms, and therefore does not need access to the actual data sets, if clustering results are available instead.

In our approach, we first compute a "co-association matrix" based on several clustering results of four publicly available yeast cell-cycle related microarray data sets. An entry (i,j) in the co-association matrix represents the fraction of partitionings in which genes i and j appeared in the same cluster. This co-association matrix is subsequently fed into a meta-level clustering algorithm, and the resulting partitioning is the "integrated" partitioning, whose biological relevance we evaluate using the Gene Ontology [35].

We are aware of the recent work of Filkov and Skiena [12] that attempts to integrate various microarray data sets through integration of the partitionings obtained by data sets clustering. They perceive the integration problem as the problem of inferring a median partitioning with respect to the given partitionings. Since the median partitioning problem is known to be NP complete [36], they propose a heuristic for computing the median partitioning, which is materialized through a simulated annealing algorithm aimed at minimizing an objective function. The quality of the "integrated" partitioning is measured by comparing the value of the objective function for the integrated partitioning with the value of the objective function computed when integrating an equal number of random partitionings of the data set. We have implemented the Filkov and Skiena's approach, and we have compared the quality of the integrated partitionings from a biological point of view in Section 5. The experimental evaluation shows that our approach produces more biologically relevant results than the "median partitioning" approach.

The rest of the chapter is organized as follows. We review the related work in Section 2. Section 3 describes the data sets used, and the preprocessing techniques we applied. Section 4 contains a brief description of the clustering algorithms used, and explains how the number of clusters for each clustering algorithm has been established. The proposed methodology for integrating a set of clustering solutions is described also in this section. Experimental results are presented in Section 5. We conclude the chapter with a discussion and directions for future work in Section 6.

2. Related Work

In this section, we review related integrated data mining techniques, followed by a discussion of the main integration approaches within the cluster ensemble formalism. With various biological data sets emerging at an increasing rate for the last few years [33], one of the major challenges faced by life sciences nowadays is the development of strategies for integrating these biological data sets, coupled with the development of integrated data mining techniques that are able to extract valuable knowledge from an integrated view of these data sources. Challenges posed by this integration from a data management point of view started to be acknowledged recently [34].

Related to our work are integrated data mining techniques that combine heterogeneous data sources in order to predict gene function. One of the first proposals for integrating various biological data sets in order to achieve increased quality in gene function prediction belongs to Marcotte *et al.* [14]. Gene expression data, phylogenetic profiles and evolutionary evidence of domain fusion are considered for determining functional relationships between yeast genes. However, pairs of genes with a functional relationship are determined separately on each data set, and then a final list with pairs of genes is compiled by concatenating the results obtained on each data set.

Pavlidis *et al.* [15] integrates gene expression data and phylogenetic profiles through the use of a support vector machine (SVM) learning algorithm in order to predict gene functional classification for yeast. The method uses simultaneously both sources of data through the means of building a heterogenous kernel which facilitates the incorporation of domain knowledge into the task at hand, i.e., weighting each data type according to its importance for classification. Although the method achieved improved functional annotation, it is difficult to extend it for integration of disparate biological data sets for gene function prediction.

Troyanskaya *et al.* [13] integrates multiple groupings of yeast genes derived from clustering of microarray data, protein-protein interactions, and pairs of genes that have experimentally determined binding sites for the same transcription factor through a Bayesian network that is constructed by experts in the field of yeast molecular biology. The network computes the probability that two genes have a functional relationship based on the evidence presented as prior probabilities, which are also assigned by domain experts. This approach has in common with the approach presented in this chapter that the input to the integration algorithm is a set of groupings of yeast genes according to some criterion of similarity. The involvement of human experts is quite substantial in Troyanskaya *et al.*'s method, making the methodology susceptible to variability in experts' opinions. Learning the network structure and priors directly from the data could potentially alleviate this effect.

Integrated data mining methods that combine and analyze heterogeneous biological data have not been developed for the purpose of predicting gene function solely, but they are also aimed at understanding protein regulation and modelling of biological networks. Recent methods that conduct an integrated analysis of diverse biological data are reviewed in [43].

Informally, the clustering problem is defined as the problem of partitioning data objects into groups (clusters) so that objects that are within the same group are similar, and objects that are in different groups are dissimilar [20]. The notion of a "cluster" has been quantified in various ways by different clustering algorithms existing in the literature. However, none of these algorithms can handle any clustering structure. In addition, given a data set, any clustering algorithm will produce a clustering solution, which in some cases may be an artifact of the clustering algorithm itself rather than a property of the data. When scarce or no prior knowledge is available regarding a dataset, choosing the "best" clustering algorithm is a hard task.

The cluster ensemble formalism proposes the integration of the information provided by multiple clustering algorithms instead of choosing the "best" clustering algorithm for a data set. The goal of integrating multiple clustering solutions is to help recover the "true" structure of the data. The assumption underlying the cluster ensemble formalism is that different algorithms will make "mistakes" in different parts of the data space, which can be "cancelled out" by aggregating multiple partitionings. This idea proved to be beneficial in the context of integration of classifier results [3, 4]. However, combining the results of multiple clustering algorithms is a more difficult task because

- The cluster labels are symbolic, and there is no explicit correspondence between the labels delivered by different clustering algorithms
- The number and the shape of the clusters in each partitioning vary based on the clustering criterion and the view of the data available to the method
- Unlike the supervised setting, there is no available ground truth to boost the overall accuracy of the ensemble

Designing a cluster ensemble requires performing the following 2 steps:

(1) Given a data set, generate a set of partitionings from which the ensemble is composed
(2) Design a strategy for combining the information embedded in the given partitionings in order to obtain a superior clustering solution

Approaches proposed for generating a set of partitionings leverage the assumption that the participating partitionings must differ to some degree, because, otherwise, no benefit can be achieved by combining identical partitionings. Various strategies have been used: (a) Exploit the stochastic nature of some clustering algorithms [17] (e.g., partitionings are obtained by running K-Means with random initialization); (b) Bagging [37] (partitionings are obtained by running the same clustering algorithm on *bootstrap* sets, which are sets of the same size as the original set of objects obtained by sampling with replacement from the original data set); (c) Random projection [38] (partitionings are obtained by running the same clustering algorithm on data sets resulted through random projection of the original data on lower dimensional spaces).

Several paradigms have been proposed for combining the information embedded in a set of partitioning.

Fred and Jain [17] introduced the "co-association matrix" approach. The integration of a set of partitionings is realized through a "co-association matrix", which is a square matrix of size equal to the number of data objects, and where an entry (i,j) counts the fraction of partitionings in which objects i and j were placed in the same cluster. The co-association matrix represents a new similarity measure for the data objects, and it is subsequently used in another clustering algorithm, which outputs the final, "integrated partitioning".

Other authors [37] have computed a consensus partitioning by solving a correspondence problem amongst the participating partitionings through heuristic re-labelling of their clusters. Some partitioning is chosen as a "reference partitioning", and all the other partitionings are re-labelled so that their agreement with the reference partitioning is maximized. The reference partitioning may or may not be a member of the ensemble. After this heuristic re-labelling is performed, an object is assigned to a cluster, based on a majority-voting procedure. This approach is restricted to partitionings with the same number of clusters, which may not be realistic in the case of gene expression data.

Strehl and Ghosh [18] introduced a hypergraph partitioning approach, where clusters in ensemble partitionings are represented as hyperedges of a hypergraph having the number of vertices equal to the number of objects in the data set. A hyperedge represents a set of objects that are located in the same cluster in one of the ensemble's partitionings. A k-way min-cut algorithm is applied on the resulted hypergraph, and each resulted connected component represents a cluster in the final, integrated partitioning. Graph-based partitioning algorithms do not perform well in cases when "natural" clusters are highly imbalanced; whether "natural" clusters are highly imbalanced or not in gene expression data is unknown.

3.　Data Preprocessing

We have chosen to illustrate our approach on gene expression data of yeast (*Saccharomyces cerevisiae*), since yeast is one of the organisms extensively studied in molecular biology, and there are a large number of resources and public microarray data sets on yeast [27].

　　We have considered four yeast microarray data sets, which are jointly known as the "cell cycle" set, which can be found at http://genome-www.stanford.edu/cellcycle/. Three of the four data sets, namely α, *cdc*15, and *elu*, were used in the study of Spellman *et al.* [1], which measured the expression levels of 6178 yeast genes under 18, 24, respectively 14 time points spanning the yeast cell cycle. The goal of this study was to identify a list of yeast genes whose transcript levels vary periodically with the cell cycle. Together with the three data sets described above, Spellman *et al.* included in their analysis a fourth data set, called *cdc*28, which also measured the expression patterns of the yeast genes under 17 time points in the cell cycle, and which was previously analyzed by Cho *et al.* [2] with the same goal as Spellman's. The names of the data sets refer to the procedures used to synchronize the yeast cultures. Yeast cells belonged to different strains, and were synchronized at different time points during the cell cycle. Gene expression levels were measured over roughly one cell cycle in the *elu* data set, two cell cycles in the α and *cdc*28 data sets, and three cell cycles in the *cdc*15 data set. All data sets contain logarithms of ratio-based measurements.

　　When measuring the expression of 6178 yeast genes during the cell cycle, it is reasonable to expect that some of the genes do not contribute to regulation. The expression patterns of these genes are "flat", i.e., they show little variation over time and the fluctuations in expression levels are more likely noise than signal. The presence of these genes can seriously hurt a clustering algorithm, because they tend to cluster together without having any biological meaning, and when using correlation-based distances, they can correlate with nearly everything. Many gene expression data analysis tools acknowledge this problem, and include within their data preprocessing toolkits several heuristics for removing genes with "flat" expression patterns [23][25].

Instead of using some heuristic for "flat" genes removal, we have used the set of 800 genes that Spellman *et al.* identified as cell cycle regulated by analyzing the four data sets described above The 800 genes can be found at http://genome-www.stanford.edu/cellcycle/. This decision is motivated by the fact that clustering cell cycle regulated genes is likely to produce clusterings that are biologically relevant, i.e., genes clustered together have similar biological function. The 800 genes that we use in this work have been identified by Spellman *et al.* based on the following procedure. Expression patterns were given an aggregate score based on a test for periodicity using a Fourier algorithm, as well as based on a correlation function that identified genes whose expression levels were similar to the expression levels of genes already known to be regulated by the cell cycle (for more details, refer to *Methods* in [1]). Genes were ranked according to the aggregate score, and a threshold aggregate score that was exceeded by 91% of known cell cycle regulated genes was computed. The selected 800 genes are the genes that had an aggregate score greater than the threshold aggregate score.

Some of the genes in the four data sets have missing expression levels. Many algorithms for gene expression analysis, such as hierarchical clustering and K-Means, require a complete matrix of gene array values as input, since they are not robust to missing data, and may lose effectiveness even with a few missing values. We have found 4.3% missing values in the set of 800 genes selected. The weighted k-Nearest Neighbor algorithm proposed by Troyanskaya *et al.* [21] is used for estimating missing values. The imputation of the missing values in a given expression pattern is done based on its k most similar expression patterns, where similarity is defined in terms of Euclidean distance. The missing values are computed as a weighted average of the similar expression patters, where the contribution of each similar expression pattern is weighted by its similarity. This method for estimation of missing values has been shown to be more robust than other commonly used methods, such as the "row average" method or filling the missing values with zeroes [21]. We have used the R package *impute* [26] that implements the method of Troyanskaya *et al.*, and we have considered k = 15 most similar expression patterns for missing value imputation.

4. Integrated Clustering

4.1. *Clustering algorithms*

Several clustering solutions are produced on the four microarray data sets by using the Pearson correlation as the distance metric, and K-Means, agglomerative hierarchical clustering, and self-organizing maps as clustering algorithms.

The selection of the clustering algorithms is motivated by the fact that these are the most widely used algorithms for clustering gene expression data.

The K-Means clustering algorithm [40] takes as input the number of clusters, K, and produces a partitioning of a data set into K disjoint clusters, so that the intra-cluster similarity is high and the inter-cluster similarity is low. Initially, K points, called "centroids" are chosen randomly in the data space, and each data object is assigned to its closest centroid. A cluster is the set of points assigned to a centroid. The centroid of each cluster is then updated as being the mean of the objects assigned to that cluster. The assignment of data objects to centroids and the update of the centroids is repeated until no changes occur or until a maximum number of iterations is reached. The K-Means algorithm minimizes the objective function given by the sum of the squared distances of each point to its closest centroid. Often, K-Means ends at a local minima of the objective function. This problem can be alleviated to some extent by using multiple random initializations. The K-Means algorithm tends to produce clusters that are as compact and separated as possible.

Agglomerative hierarchical clustering [32] is a bottom-up approach that starts by placing each object in its own cluster, and then, at each step, merges the most two similar clusters into a single cluster, until all objects are in a single cluster or a certain termination condition has been met. There are several variations of agglomerative hierarchical clustering based on the definition of similarity used to merge two clusters. Based on findings reported in previous work [16], the "complete linkage" variant of the hierarchical clustering algorithm is used. In the "complete linkage" variant, a cluster is represented by all its points, and the distance between two clusters is the distance between the *farthest* two points in the two clusters. The outcome of an agglomerative hierarchical clustering is a *dendrogram*, i.e., a tree that shows the order in which clusters are merged. Any hierarchical clustering algorithm does not assume a predefined number of clusters. Instead, a particular number of clusters is obtained by "cutting" the dendrogram at a certain height. Note that the result of a hierarchical clustering algorithm is deterministic for a specified K.

Self-organizing maps (SOB) were introduced by [22] and used the first time to analyze gene expression data by [[7]. A self-organizing map consists of a set of neurons that are arranged according to a specified topology. Most often, the neurons are arranged within a grid. Conceptually, a neuron corresponds to the centroid of a cluster. Initially, the neurons are initialized at random positions in the data space. The genes to be clustered are processed in random order as follows: given a gene, the neuron that is most similar to this gene is modified, so that it more closely resembles the expression pattern of that gene. Also, the neurons that are neighbors of this neuron in the grid are also updated so that they too resemble the gene's expression pattern more closely. This procedure is iterated several times until no significant changes in neurons are registered. At the end, the neurons have converged to the cluster's centroids, and neighboring centroids correspond to clusters with genes having similar expression patterns. According to [16], in the case of gene expression data, the most "square-like" layout for the self-organizing maps is preferred to other layouts, i.e., a layout which minimizes the ratio of the perimeter length to the area.

Pearson correlation distance metric is preferred to the Euclidean distance for gene expression data clustering [6] due to the fact that it is invariant under scalar transformation of the data. Given two genes g_1 and g_2, whose expression vectors are $x_1,...,x_n$, and $y_1,...,y_n$, respectively, where n is the number of samples measured, the Pearson correlation distance metric for g_1 and g_2 is given by:

$$d = 1 - \frac{1}{n} \sum_{i=1}^{n} \frac{(x_i - \overline{x}) * (y_i - \overline{y})}{\sigma_x * \sigma_y}$$

where $\overline{x}$ and σ_x are the average, and the standard deviation, respectively, of the numbers $x_1, ..., x_n$, and $\overline{y}$ and σ_y are the average, and the standard deviation, respectively, of the numbers $y_1, ..., y_n$.

We note that all clusterings are produced with the R package [R]. Deciding the number of clusters that each partitioning should have is a difficult task when the underlying distribution of the data is unknown. For each (data set_clustering algorithm) pair, the number of clusters that maximizes a certain measure of clustering quality is chosen. We have chosen the silhouette coefficient [31] to measure the quality of a clustering result.

For each gene g, let $a(g)$ be the average distance of g to all other genes of the cluster to which g belongs. For all other clusters C, let $d(g,C)$ be the average distance of g to all genes of C. Let $b(g) := \min_C d(g,c)$ be the average distance of g to all genes in its "neighbor" cluster, i.e., the closest cluster to which it does not belong. The silhouette width of gene g is defined as $s(g) := (b(g) - a(g) / \max(a(g), b(g))$. The silhouette width takes values between 0 and 1, the semantics being that a gene with $s(g)$ close to 1 is well clustered, while an $s(g)$ around 0 signifies that the gene lies between two clusters, and an negative $s(g)$ indicates that the gene is probably placed in the wrong cluster. The silhouette coefficient of a clustering is defined as the average of the silhouette widths of all genes, and the higher the silhouette coefficient of a clustering, the higher the quality of the clustering.

Table 1 summarizes the number of clusters found for each (data set_clustering algorithm) pair by computing the silhouette coefficient for multiple runs of the clustering algorithm with different values for k, the number of clusters, and by selecting the k for which the largest silhouette coefficient was obtained. In each case, values for k in the set

Table 1. Number clusters for each partitioning.

Data set_Clsutering Algorithms	Number Clusters
a_KM	50
a_H	50
a_S	20
c15_KM	20
c15_H	20
c15_S	20
c28_KM	20
C28_H	20
C28_S	20
E_KM	20
E_H	20
E_s	20

{10,20,30,40,50,60,70,80,90,100} were tested, based on the assumption that each cluster should have in average more than a minimum number of genes. Clustering the four data sets by three different clustering algorithms yields 12 partitionings, coded as a (data set_clustering algorithm) pair. In our notation, a, $c15$, $c28$, and e represent the four data sets α, $cdc15$, $cdc28$, respectively elu, as described in Section 3, and the abbreviation KM stands for the K-Means clustering algorithm, H stands for hierarchical clustering, and S stands for self-organizing maps.

Interestingly, the number of clusters as computed by the methodology described above is the same for many of the (data set_clustering algorithm) pairs. This may be a consequence of the fact that the same set of genes measured under similar experimental conditions is clustered. Note that having the same number of clusters in each partitioning is not required by the integration methodology that is described below.

The K-Means algorithm may often converge to a local optimum; therefore each partitioning obtained with K-Means was selected by running K-Means three times with random initial configuration, and keeping the clustering solution with the highest silhouette coefficient. A similar procedure was applied to the partitionings obtained with self-organizing maps, i.e., the order of the genes expression patterns is randomly permuted, since this algorithm is sensitive to the ordering of the objects to be clustered.

We are interested in measuring the agreement in clustering genes amongst the 12 partitionings. Many measures have been proposed for assessing the similarity between two partitionings of the same data set. The most often used similarity measure is the Rand index [41], which counts the number of agreements in two partitionings. Given two partitionings P_1 and P_2 of a set of n objects, the Rand index is defined as

$$Rand(P_1, P_2) = (a+d)/\binom{n}{2},$$ where a is number of object pairs co-

clustered in both partitionings; b is the number of object pairs co-clustered in P_1, but not in P_2 ; c is the number of object pairs co-clustered in P_2 , but not in P_1; and d is the number of object pairs not co-clustered

in both partitionings. Clearly, $a+b+c+d = \binom{n}{2}$.

However, the Rand index is not adjusted for chance, meaning that two independent random partitionings may have a Rand index score of 1 on the average. We have used a refined version of the Rand index, namely the Adjusted Rand index [42], whose formula is:

$$AdjustedRand(P_1, P_2) = (a+d-n_c)/(\binom{n}{2} - n_c),$$ where n_c is a corrective

factor that can be computed as $n_c = ((a+b)(a+c) + (c+d)(b+d))/\binom{n}{2}.$

Same measure of agreement between two partitionings have been used in the method of Filkov and Skiena, to which we want to compare our method.

Two partitionings with an Adjusted Rand index of 1 show complete agreement, while an Adjusted Rand index of 0 indicates total disagreement. Table 2 shows the pair-wise Adjusted Rand index amongst the 12 partitionings that were produced. The Adjusted Rand index takes values ranging between 0.79 and 0.95, suggesting in general good agreement amongst the partitionings, but not a perfect consensus.

Table 2. Adjusted Rand index.

	a_H	a_KM	a_S	c15_H	c_KM	c15_S	c28_H	c28_KM	c28_S	e_H	e_KM	e_S
a_H	1	0.95	0.93	0.91	0.9	0.85	0.9	0.91	0.88	0.86	0.91	0.88
a_KM		1	0.94	0.91	0.91	0.85	0.9	0.92	0.88	0.87	0.92	0.88
a_S			1	0.91	0.91	0.85	0.89	0.91	0.88	0.87	0.91	0.88
c15_H				1	0.91	0.83	0.87	0.89	0.85	0.84	0.88	0.85
c_KM					1	0.83	0.87	0.89	0.85	0.84	0.88	0.85
c15_S						1	0.81	0.83	0.81	0.79	0.83	0.81
c28_H							1	0.9	0.84	0.84	0.87	0.84
c28_KM								1	0.86	0.85	0.89	0.85
c28_S									1	0.81	0.86	0.84
e_H										1	0.87	0.81
e_KM											1	0.85
e_S												1

Notice: The abbreviations *a*, *c*15, *c*28, and *elu* stand for the four data sets α, *cdc*15, *cdc*28, respectively *elu*. *KM* stands for the K-Means clustering algorithm, *H* stands for hierarchical clustering, and *S* stands for self-organizing maps, respectively. An Adjusted Rand index of 1 indicates complete agreement between the corresponding partitionings, while an Adjusted Rand index of *0* indicates total disagreement.

4.2. *Integration methodology*

Formally, let $G = \{G_1, G_2, ..., G_n\}$ be a set of genes that are measured in $M_1, M_2, ..., M_m$ microarray experiments. The microarray experiments are similar in the sense that they measure the same biological process, such as cell cycle, heat shock response, etc. Clustering the genes in each microarray experiment yields a set of partitionings $\{P_1, P_2, ..., P_l\}(l \geq m)$ of the set of genes G, where each partitioning is a set of disjoint, exhaustive and non-empty clusters.

The idea underlying the integration methodology is the "majority voting" concept, which postulates that the judgement of a group is superior to those of individuals. Bagging (e.g., [3]) and boosting (e.g., [4]), two well known techniques for combining the outcome of multiple classifiers, are based on this concept. The intuition behind integrating a set of partitionings through a majority voting approach is that objects belonging to a "natural" cluster are likely to be co-located in the same cluster by different clustering algorithms. The integrated partitioning is expected to have a compensatory effect by levelling out the noise existing in each of the participating partitionings.

Integrating the knowledge embedded within each of the partitionings $\{P_1, P_2, ..., P_l\}$ is achieved through a "co-association matrix" approach [17]. The "co-association matrix" is a similarity matrix for the genes in G, where an entry (i,j) represents the fraction of partitionings $\{P_1, P_2, ..., P_l\}$ in which genes i and j were clustered together. The underlying idea of this mechanism is that genes belonging to the same "natural" cluster are likely to be co-located in the same cluster by different clusterings under slightly different experimental conditions. The entries in the matrix are between 0 and 1, with 0 meaning that the corresponding genes were never co-clustered in any of the partitionings, while 1 indicates that the corresponding genes were co-clustered in all partitionings.

The co-association matrix is transformed into a distance matrix by subtracting from 1 the value of each entry. This distance matrix serves as input to a hierarchical clustering algorithm (with complete linkage). The resulted dendrogram is cut into a number of clusters for which a maximal value of the silhouette coefficient is obtained. The outcome of the integration procedure is a new partitioning of the set of genes.

5. Experimental Evaluation

5.1. *Evaluation methodology*

Evaluating the results of a clustering algorithm is a difficult and often ill-posed task [20]. The proposed integration methodology does not generate any model for the integrated partitioning, thus internal evaluation criteria are less appropriate in this case. In addition, the biological relevance is the most important criterion in assessing the quality of a given partitioning of gene expression data.

Given a partitioning of a set of genes into a set of clusters, the quality of the partitioning is proportional to its capability of grouping together genes having similar function. For any pair of genes that are co-clustered in the given partitioning, a methodology based on the Gene Ontology [35] is used to determine whether the genes have similar biological function. A pair of co-clustered genes that have similar biological function is counted as a "true positive" *(TP)*; otherwise is counted as a "false positive" *(FP)*. The quality or "accuracy" of the given partitioning is computed as the fraction of true positives pairs out of the total number of pairs that were co-clustered, i.e., as the fraction *TP/(TP+FP)*.

The Gene Ontology is used for assessing whether two genes have similar biological function. GO is a structured terminology that describes genes and genes products, structured as a directed acyclic graph (DAG). The relationship between terms is a parent-child relationship, where the parents of any term are less specific than the child. GO is divided in three separate ontologies, one for cellular component (CC), one for molecular function (MF) and one for biological process (BP). There is a single root for all ontologies as well as separate root nodes for each of the three ontologies. We have considered only the biological process ontology, which is the most relevant of the ontologies for evaluating groupings of genes based on functional relationships [13].

GO is only a collection of terms; the annotation of yeast genes with GO terms is produced by the Saccharomyces Genome Database at Stanford [28]. A gene is annotated only with the most specific GO terms that are applicable to it, which may be to restrictive for the evaluation strategy. Thus, the annotation of a given gene is enriched with all ancestors terms of the most specific terms associated with it up to the level three of the biological process ontology.

Two genes are considered to have similar function if the intersection of their annotations is not empty, except the case when the intersection consists only of the GO term "GO: 0000004", which stands for "biological process unknown". Note that "true positive" and "false positive" pairs are counted only when both genes in the pair have GO annotations.

The BioConductor's packages YEAST and GO have been used in order to annotate the yeast genes with GO terms as described above (Bioconductor is an open source software project for the analysis and comprehension of genomic data comprised of a collection of *R* packages

[BioConductor]. A similar evaluation procedure was conducted by Troyanskaya *et al.* in [13].

5.2. Results

The proposed integration methodology (Section 4.2) was applied to the 12 partitionings described in Section 4.1. The resulted integrated partitioning is referred to as the "co-association partitioning", and abbreviated by "CP". The same 12 partitionings were also integrated based on the strategy of Filkov and Skiena [12], and the resulted integrated partitioning is referred to as the "median partitioning", and abbreviated by "MP". The accuracies of each of the individual partitionings, as well as the accuracies of the co-association and median partitionings are computed. The results are presented in Fig. 1.

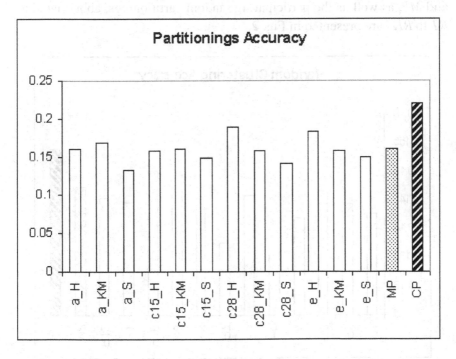

Figure 1: Partitionings accuracy.

The co-association partitioning (*CP*) achieves an improvement in quality of 65% over the individual partitioning with the lowest quality, and an improvement of 16% over the individual partitioning with the highest quality. The median partitioning (*MP*) is not able to improve over the best individual partitioning. The co-association partitioning achieves an improvement of 36% in quality over the median partitioning.

It is interesting investigating the performance of the integration methodology on randomly generated partitionings, to see whether clustering results of gene expressions are in fact better reflected by GO annotations than random partitionings. 12 random partitionings of the 800 genes are generated, where each partitioning has a random number of clusters between 10 and 100, and each gene is assigned randomly to one of these clusters. The random partitionings are integrated through both the "co-association matrix", as well as the "median partitioning" approaches. The accuracies of the resulting integrated partitionings, *CP* and *MP*, as well as the participating random partitionings, abbreviated as *R1* to *R12*, are presented in Fig. 2.

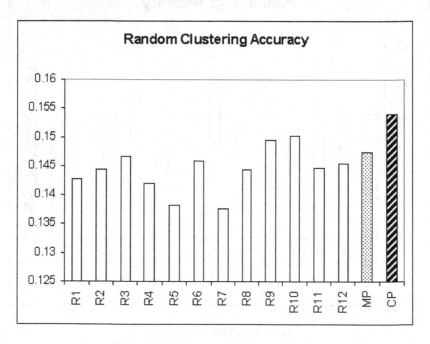

Figure 2: Random partitionings accuracy.

We observe that the accuracies of all algorithms on random partitionings are noticeably lower than for clustering results on microarray data. This is consistent with our expectations, since in randomly partitioned data, the clusters are random, are thus are not expected to have high GO similarities. To quantify this observation, we test whether the results of the co-association approach on sets of partitionings derived with the above-listed clustering algorithms are significantly better than the results of the co-association approach on sets of random partitionings, using the Wilcoxon sum rank test [5]. We generate an equal number of sets of random partitionings of the data set and sets of partitionings computed with the three clustering algorithms of choice, where K-Means is run with different random initializations, and self-organizing maps are run with random presentation of the objects to be clustered. The Wilcoxon sum rank test indicates that the accuracies of the integrated partitionings computed on sets of partitionings derived with the above-listed clustering algorithms are significantly better (at a level of significance $\alpha = 0.05$ than the accuracies of the integrated partitionings computed on sets of random partitionings.

5.3. *Discussion*

Inspecting the results presented in the previous section reveals an interesting observation. Although the clustering results are significantly better than random partitionings, the absolute accuracy values are not very high. The cause for this fact is rooted in the evaluation strategy based on the Gene Ontology. The evaluation of a clustering algorithm on gene expression data based on the Gene Ontology is not without problems, as also acknowledged by other authors [16]. As pointed out by previous work [13], GO may contain annotation errors; the function of many genes is not determined yet, and there might be some biases in the set of genes currently lacking annotation. However, it is unlikely that any of these problems is differentially affecting the evaluation of any of the partitionings, and since GO represents the current knowledge regarding biological functions of genes, it provides a reasonable evaluation framework for gene clusterings, which we believe measures the relative performances of the studied approaches in a reliable way.

6. Conclusions

A different approach for clustering of gene expression data based on the integration of clusterings of multiple microarray experiments measuring the expression of a given organism's genes under similar experimental conditions is presented. The integration strategy is a "majority voting" approach based on the assumption that objects belonging to a "natural" cluster are likely to be co-located in the same cluster by different clustering algorithms. The evaluation strategy based on the Gene Ontology reveals the benefit of integration, as well as the superiority in terms of biological relevance of the presented integration approach to a previously proposed integration approach. The evaluation strategy also indicates some of the caveats involved in using the Gene Ontology for measuring the quality of clusterings of gene expression data.

Our results indicate that in principle there is a great potential for improving the knowledge extracted from clusters in microarray data obtained by individual studies, without the necessity to share the actual data. Instead, only the evidence published as clustering results can be integrated to acquire biologically more meaningful results that contained in any of the individual results.

For future work, we plan to investigate various integration methodologies from a theoretical perspective in order to develop adequate objective functions that would allow us to estimate the benefit of the integration independently of the Gene Ontology.

References

[1] Spellman, P. T. and Sherlock, G. and Zhang, M. Q. and Iyer, V. R. and Anders, K. and Eisen, M. B. and Brown, P. O. and Botstein, D. and Futcher, B. "Comprehensive Identification of Cell Cycle-regulated Genes of the Yeast Saccharomyces cerevisiae by Microarray Hibridization," Molecular Biology of the Cell, 9, pp.3273-3297, 1998.

[2] Cho, R. J. and Campbell, M. J. and Winzeler, E. A. and Steinmetz, L. and Conway, A. and Wodicka, L. and Wolfsberg, T. G. and Gabrielian, A. E. and Landsman, D. and Lockhart, D. and Davis, R. W. "A Genome-Wide Transcriptional Analysis of the Mitotic Cell Cycle," Molecular Cell, 2, pp.65-73, 1998.

[3] Breiman, L. "Bagging Predictors," Machine Learning, 24, pp.123—140, 1996.

[4] Freund, Y. and Schapire, R. E. "Experiments with a new boosting algorithm," Proceedings of International Conference on Machine Learning (ICML), 1996.

[5] Wilcoxon, F. "Individual comparisons by ranking methods," Biometrics, 1, pp.80—83, 1945.

[6] Eisen, M. B. and Spellman, P.T. and Brown, P. O. and Botstein, D., "Cluster analysis and display of genome-wide expression patterns," Proc. Natl. Acad. Sci. USA, 85, pp.14683—16868, 1998.

[7] Tamayo, P. and Slonim, D. and Mesirov, J. and Zhu, Q. and Kitareewan, S. and Dmitrovsky E. and Lander, E. and Golub, T. R. "Interpreting patterns of gene expression with self-organizing maps: methods and application to hematopoietic differentiation," Proc. Natl. Acad. Sci. USA, 85, pp.2907—2912, 1999.

[8] Tavazoie, S. and Hughes, J. D and Campbell, M. J. and Cho, R. J. and Church, G. M., "Systematic determination of genetic network architecture," Nature genetics, 22, pp.281—285, 1999.

[9] Brown, M. and Grundy, W.N. and Lin, D. and Cristianini, N. and Sugnet, C.W. and Furey, T.S. and Ares, M.J. and Haussler, D., "Knowledge-based analysis of microarray gene expression data by using support vector machines," Proc. Natl. Acad. Sci USA, 97, pp.262—267, 2000.

[10] Golub, T. R. and Slonim, D.K. and Tamayo, P. and Huard, C. and Gaasenbeek, M. and Mesirov, J. P. and Coller, H. and Loh, M. L. and Downing, J. R. and Caligiuri, M. A. and Bloomfiled, C. D. and Lander, E. S., "Molecular classification of cancer: class discovery and class prediction by gene expression monitoring," Science, 286, pp.531—537, 1999.

[11] Zhang, H. and Tu, C. Y. and Singer, B. and Xiong, M., "Recursive partitioning for tumor classification with gene expression microarray data," Proc. Natl. Acad. Sci. USA, 12, pp.6730—6735, 2001.

[12] Filkov, V. and Skiena, S., "Integrating Microarray Data by Consensus Clustering," International Journal on Artificial Intelligence Tools, pp.20—42, 2004.

[13] Troyanskaya, O. G. and Dolinski, K. and Owen, A. B. and Altman, R. B. and Botstein, D. (2003), "A Bayesian framework for combining heterogeneous data sources for gene function prediction in Saccharomyces cerevisiae," it Proc. Natl. Acad. Sci USA, 100, pp.8348—8353, 2003.

[14] Marcotte, E. M. and Pellegrini, M. and Thompson, M. J. and Yeates, T. O. and Eisenberg, D., "A combined algorithm for genome-wide prediction of protein function," Nature, 402, pp.83—86, 1999.

[15] Pavlidis, P. and Weston, J. and Cai, J. and Noble, W. S., "Learning Gene Funcional Classifications from Multiple Data Types," Journal of Computational Biology, 9, pp.401—411, 2002.

[16] Gibbons, F. D. and Roth, F. R., "Judging the Quality of Gene Expression-Based Clustering Methods Using Gene Annotation," Genome Research, pp.1574—1581, 2002.

[17] Fred, A. L. N. and Jain, A. K., "Data Clustering using Evidence Accumulation," Proceedings of the International Conference on Pattern Recognition (ICPR), 2002.

[18] Strehl, A. and Ghosh, J., "Cluster ensembles - a knowledge reuse framework for combining multiple partitions," Machine Learning Research, 3, pp.583—617, 2002.

[19] Topchy, A. and Jain, A. K. and Punch, W., "Combining multiple weak clusterings," Proceedings of the International Conference on Data Mining (ICDM), 2003.

[20] Jain, A. K. and Dubes, R. C. (1998). Algorithms for Clustering Data, Prentice Hall.

[21] Troyanskaya, O. and Cantor, M. and Sherlock, G. and Brown, P. and Hastie, T. and Tibshirani, R. and Botstein, D. and Altman, R. B. "Missing value estimation methods for DNA microarrays," Bioinformatics, 17, pp.520—525, 2001.

[22] Kohonen, T. "Self-organized formation of topologically correct feature maps," Biological Cybernetics, 43, pp.59—69, 1982.

[23] Cluster and Tree View. http://rana.lbl.gov/EisenSoftware.htm.

[24] EPCLUST. http://ep.ebi.ac.uk/EP/.

[25] GEPAS. http://gepas.bioinfo.cnio.es/.

[26] R package. http://www.r-project.org/.

[27] Saccharomyces Genome Database (SGD).
http://genome-www.stanford.edu/Saccharomyces/.

[28] Stanford Microarray Database (SMD). http://genome-www5.stanford.edu/.

[29] Critical Assessment of Microarray Data Analysis (CAMDA).
http://www.camda.duke.edu/camda03/.

[30] BioConductor. http://www.bioconductor.org/.

[31] Rousseeuw, P.J., "Silhouettes: A graphical aid to the interpretation and validation of cluster analysis," J. Comput. Appl. Math., 20, pp.53—65, 1987.

[32] Kaufman, L. and Rousseeuw, P.J. (1990). Finding groups in data: an introduction to cluster analysis. New York: John Wiley & Sons.

[33] Baxevanis, A. D., "The Molecular Biology Database Collection: 2003 Update," Nucleic Acids Research, 31, pp.~1—12, 2003.

[34] Hernandez, T. and Kambhampati, S., "Integration of Biological Sources: Current Systems and Challenges Ahead," Sigmod Record, 2004.

[35] Ashburner, M. C. and Ball, C. A. and Blake, J. A. and Botstein, D. and Butler, H. and Cherry, J. M. and Davis, A. P. and Dolinski, K and Dwight, D. and Eppig, J. T. and Harris, M. A. and Hill, D. P. and Issel-Tarver, L. and Kasarskis, A. and Lewis, S. and Matese, J. C. and Richardson, J. E. and Ringwald, M. and Rubin, G. M. and Sherlock, G., "Gene ontology: tool for the unification of biology," Nature Genetics, 25, pp.25—29, 2000.

[36] Barthelemy, J. P. and Leclerc, B., "The median procedure for partition," DIMACS Series in Discrete Mathematics, 19, pp.3—34, 1995.

[37] Dudoit, S. and Fridlyand, J., "Bagging to improve the accuracy of a clustering procedure," Bioinformatics, 19, pp.1090—1099, 2003.

[38] Fern, X. Z. and Brodley, C. E, "Random projection for high dimensional data clustering: a cluster ensemble approach," In Proceedings of the International Conference on Machine Learning (ICML), 2003.
[39] Kellam, P. and Liu, X. and Martin, J. and Orengo, C. and Swift, S. and Tucker, A. , "Comparing, constrasting, and combining clusters in viral gene expression data," In Proceedings of the 6th Workshop on Intelligent Data Analysis in Medicine and Pharmacology, 2001.
[40] MacQueen, J., "Some methods for classification and analysis of multivariate observations," Proceedings 5th Berkeley Symp. Math. Statist, Prob.,1, pp.281—297, 1967.
[41] Rand, W., "Objective criteria for the evaluation of the clustering methods," Journal of the American Statistical Association, 66, pp.846—850, 1971.
[42] Hubert, L. and Arabie, P., "Comparing partitions," Journal of Classification, 2, pp.193—218, 1985.
[43] Troyanskaya, O. G., "Putting microarrays in a context: Integrated analysis of diverse biological data," Briefings in Bioinformatics, 6 (1), pp.34—43, 2005.

CHAPTER 5

COMPLEXITY AND SYNCHRONIZATION OF EEG WITH PARAMETRIC MODELING

Xiaoli Li

CERCIA, School of Computer Science, The University of Birmingham, Edgbaston, Birmingham, B15 2TT, UK

E-mail: xiaoli.avh@gmail.com

Abstract

EEG recordings are often applied in the neuroscience research and the diagnosis of neural disorder. So far, EEG data analysis still is one of the vital topics in the neuroscience this is because that the series analysis of EEG could be used to understand the complex behavior of neural networks. We firstly review the analysis method of EEG recordings, it is found that the complexity and synchronization of the EEG recordings are recently paid more attentions because these definitions are more helpful to understand the dynamics of neural networks than others. In this chapter a simple parametric modeling of EEG series is introduced to compute the complexity and synchronization of EEG data sets. The parametric modeling is expected to measure the complexity and synchronization of EEG data sets, except that it has been widely applied to obtain the features and spectrum of EEG data. First of all, a timeb varying auto-regression (AR) based on a finite sample information criterion is applied to calculate an order of parametric modeling, i.e. complexity measure, by a window technique; this measure can efficiently track a dynamical change of EEG data, such as detection of the epileptic seizures in EEG. Then, a time varying multivariate AR model is applied to measure the synchronization of multi-channel EEG signals; the test results show that the synchronization of EEG data can identify the difference between the normal and seizure states of epilepsy patients. Case studies indicate that the complexity and synchronization measures of the EEG data sets based on parametric modeling could be used to diagnose epilepsy or analyze EEG dynamics.

1. Introduction

1.1. *Brief review of EEG recording analysis*

The interaction between the cellular and synaptic mechanisms is one of the most fundamental problems in neuroscience [Traub *et al.*, 1998], it has a direct relation to brain functions (e.g. memory and learning) [e.g. Ward, 2003] and pathological mechanisms (e.g. epilepsy) [Traub *et al.*, 1999]. Neuronal oscillation represents a general feature of neural firing patterns, which is produced by dynamical interplay between cellular and synaptic mechanisms [Wang, 2003]. Thus, the analysis of neuronal oscillations (EEG recordings) provides critical insights into cellular and synaptic mechanisms. Previous findings show the neuronal oscillations temporally link neurons into assemblies, facilitate synaptic plasticity, and support temporal representation and long-term consolidation of information [Buzsaki, 2004]. The synchronous activity of oscillating networks can be viewed as the critical 'middle ground' linking signal-neuron activity to behaviour [e.g. Engel *et al.*, 2001; Traub *et al.*, 1999; Buzsaki *et al.*, 2004].

As can be shown in Fig. 1, many computational methods have been proposed to analyze EEG recording. These methods can be classified as linear and nonlinear methods [Li, 2003, 2004] or one, two and multiple dimensions EEG analysis methods. (i) *Linear method with Fourier transforms and statistics*. This transform has a limitation in analyzing neuronal oscillations; this is due to the fact that neuronal oscillations are limit-cycle and weakly chaotic oscillators and share features of both harmonic and relaxation oscillations [Glass, 2001; Li *et al.*, 2004]. (ii) *Time frequency methods*. To describe the temporal information of neuronal oscillations, time-frequency analysis has been proposed to analyze neuronal oscillations [Li, 2004]. Further work is necessary to develop new methods based on time frequency space to directly describe the interaction between the cellular and synaptic mechanisms. (iii) *Nonlinear methods with Chaos.* These methods [e.g. Babloyantz *et al.*, 1985] still need improvement, including: hybridization with linear methods, dependence on length of EEG recording, and robustness in the presence of high level noise. (iv) *Synchronization.* Synchronous neuronal oscillations can reveal much about the origin and nature of cognitive processes (e.g. memory, attention and consciousness) and pathological processes (e.g. epilepsy). Conventional methods of analyzing synchronization include cross-correlation, mutual information, spectrum based coherence and nonlinear interdependence [e.g. Mormann *et al.*, 2000; Carmeli *et al.*, 2005; Le Van Quyen *et al.*, 2005]. The limitations of these methods include the loss of temporal information of neuronal oscillations and interference from noise. (v) *Phase synchronization.* Phase synchronization measure has become a main method in analyzing the interaction between brain regions [Varela *et al.*, 2001]. Present methods of phase synchronization have at least two limitations: the instantaneous phase computation of neuronal oscillation and statistic analysis of phase synchronization. In addition, the coupling direction calculation of neuronal oscillation is still omitted in neuroscience applications. (vi) *Analysis and visualization of multiple EEGs.* The synchronization, similarity, and complexity of multiple neuronal oscillation recordings and their visualization (3D topography) are a new challenge for us at present. More details on the EEG analysis can be found in the prefect review paper [Thakor and Tong, 2004]. It is noticeable that the complexity and synchronization measure of EEG recordings are paid more attentions [e.g. Jouny *et al.*, 2005; Tononi *et al.*, 1998; Rezek *et al.*, 1998; Ferrillo *et al.*, 2000]. In this chapter, we will try to make use of a simple parametric modeling to tackle this interesting issue in the EEG data analysis.

1.2. *AR modeling based EEG analysis*

Autoregression models (AR) and time varying autoregression (TVAR) have been applied to analyze EEG signals. In light of recent publications, the method mainly focuses on the following topics: (1) features extractions; (2) spectral analysis; (3) complexity estimation; (4) decomposition of an EEG signal; (5) dynamic characteristic analysis; and (6) synchronization measure of multi-channels EEG. The first three themes belong to the classical application of AR and TVAR model, especially the spectral analysis of EEG signals. The last three themes are new applications of AR and TVAR models to EEG signals.

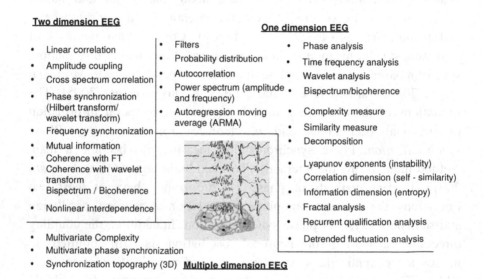

Figure 1: Overview of analysis methods for EEG recordings.

In the practical application, a notstationary process can often be divided into many stationary processes. Thus, based on this point, the AR and TVAR can be applied to analyze an EEG signal. Case study in [Tseng *et al.*, 1995] can confirm this point, i.e. an AR model with the order 8 can efficiently represent the 96% of 900 EEG segments that contain eight classes. The parameters of an AR model often are used as the features of an EEG signal for classification and prediction [Sharma *et al.*, 1997; Kong *et al.*, 1997]. AR model based spectral analysis of an EEG signal can provide better result than FFT method [Kiymik *et al.*, 2001] because the estimated spectrum of an EEG signal based on AR model is a continuous frequency function. The other application of AR model is to estimate the complexity of an EEG signal. The basic idea is how many coefficients are needed to approximate an EEG signal, the signal may be called more complex than one which requires only a few. In Rezek *et al.* (1998), the TVAR model order is taken as the complexity estimation of a system.

Except the above applications, recently some new methods based on AR or TVAR models are proposed, including decomposition, dynamic characteristics analysis and synchronization measure. In Krystal *et al.* (2000), a new analysis technique with TVAR model is proposed to decompose an EEG signal into a set of latent components with time – varying frequency content. In Anderson *et al.* (1998), multivariate AR models are also proposed to analyze the EEG signals for classification of spontaneous EEG signals during mental tasks. Synchronization of multi-channel EEG signals is very important characteristic of brain activity, and that is very helpful to analyze the brain disorder. Synchronization measure of multi-channel ictal and interictal EEG signals is presented in Franaszczuk *et al.* (1999), which is based on the residual covariance matrix of a multi-channel AR model. In Li (2003), a calculation of the damping time of an EEG based on AR model is proposed to identify the changes of normal condition and epileptic seizures of brain.

In this chapter, we further explore the time varying AR model to estimate the complexity and synchronization of an EEG signal for identifying seizures. The basic idea of complexity measure is based on following principle, namely more many coefficients are needed to approximate an EEG signal, the system is very complex; otherwise it is a simple system. Synchronization of multi-channel EEG signals is one of the mort important characteristics of brain activity. Many researchers have proved that synchronization of multiple EEG signals is very helpful to diagnose the brain disorder [Mormann *et al.*, 2000]. This chapter will address some details on the application of the TVAR model to the EEG signals, focus on complexity and synchronization measures. In the following section, the AR and TVAR model are reviewed. The complexity estimation to EEG signal are addressed in Section 3. In Section IV, a new method for synchronization measure of multi-channel ictal and interictal EEG signals is introduced. The conclusion is given in the final section.

2. TVAR Modeling

Assumes that the value of the current sample, $y(t)$, in a data sequence, $y(1), y(2), ..., y(N)$, can be predicted as a linearly weighted sum of the p most sample values, $y(t-1), y(t-2),..., y(t-p)$, where p is the model order and is generally chosen to be much smaller than the sequence length, N. The predicted values of $y(t)$ can be expressed as follows:

$$\tilde{y}(t) = -\sum_{i=1}^{p} a_i y(t-i) \tag{5.1}$$

where weight a_i denote the ith coefficient of the pth-order model. The two most popular algorithms Levinson-Durbin algorithm and Burg algorithm [Stoica *et al.*, 1997], can be used to estimate the coefficients of AR model.

TVAR model has been proposed to treat with time varying (non-stationary) data; the parameters of TVAR modeling vary with time, in contrast the parameters of AR model is fixed. Adaptive algorithm has a very useful function to insure TVAR model for online and real-time applications, such as Kalman filtering [Haykin, 1996], Recursive Least Squares (RLS) [Patomaki *et al.*, 1995], the Least Mean Squares (LMS) algorithms and recursive AR algorithms [Akay, 1994]. A TVAR model with order p can be written as

$$y(t) = a_1(t)y(t-1) + a_2(t)y(t-2) + ... + a_p(t)y(t-p) + x(t)$$
$$= \mathbf{a}(t)\mathbf{Y}(t-1) + x(t) \qquad i = 1,2,...,p \tag{5.2}$$

Clearly, the model coefficients $a_i(t)$in the TVAR model are time varying. In the ideal case, x(t) is a purely random process with zero mean and variance σ_x^2, and the cross covariance function is $E\{y(t)x(t+k)\}$ because $x(t)$ is uncorrelated with the signal y(t). TVAR model parameters can be estimated using a variety of adaptive algorithms. In this study, the predictive least squares algorithm and Kalman filtering in [Schoegl *et al.*, 1997, 2000; Sharma *et al.*, 1997; Stoica *et al.*, 1997] are employed.

An issue that is of central importance to the successfully application of AR model is the selection of an appropriate value for the model order, p. The correct model order for a given data sequence is not known in advance, it is desirable to minimize the model's computational complexity by choosing the minimum value of p that adequately represents the signal being modelled. The method based on the asymptotical theory is one of the most important methods for the AR estimation algorithms [Akaike, 1974]. For finite samples, estimated parameters, reflection coefficients and residuals depend on the order estimation of AR model [Broersen *et al.*, 1993], however these differences are not accounted for in the asymptotical theory. So the second category of order selection criteria for finite samples is proposed, i.e. the predictive least squares (PLS) criterion [Rissanen, 1986]. The selection with PLS is based on the assumption that the true or best order is a constant and independent of the sample size used for estimation. Unfortunately, in fact the best AR order depends on the sample size. Hence, the best order for "honest" predictions depends on the variable number of observations that is used to estimate the parameters in PLS algorithms and is not always a constant [Broersen, 2000]. This limits the processes for which the PLS criterion can be used. The third category called the finite sample information criterion (FSIC) is proposed [Broersen, 1998], the performance of FSIC remains the best for all estimation methods. And the results of order selection are independent of the choice of the maximum candidates order. Another criterion for the third category is the Combined Information Criterion (CIC), which combines the favorable asymptotical penalty factor 3 with the increased resistance against the selection of too high model order in finite samples [Broersen, 1993]. Simulations show that the expected performance of FSIC and CIC is realized for different AR estimation methods and for various sample sizes [Broersen, 2000].

3. Complexity Measure

When the parts or elements in a system but involved in various degrees of subordination, not simply coordinated, the system is complex. In physics, complexity is an attribute that is often employed generically to describe dynamics of a system. An acceptable notion of complexity is that systems are neither completely regular nor completely random. In term of the description of complexity, the arrangement and interactions of neurons in a brain is obviously extremely complex. Iasemidis and Sackellares [Iasemidis and Sackellares, 1996] suggested that the onset of a seizure represents a spatiotemporal transition from a complex to a less complex (more ordered) state. [Lehnertz and Elger, 1997] demonstrated decreased neuronal complexity in the primary epileptogenic area in preictal state. This is true that complexity of EEG signals is correlated with the patients' condition.

Direct assessment of EEG signals complexity offers certain advantages in clinical research. Complexity of EEG signals not only is an intuitive description, also makes the comparison between patients possible, as the complexity is insensitive to absolute measures (frequency or amplitude). From now on, there exist some methods to quantify signal complexity. First method is the spectral complexity of a time series, called spectral entropy, which has been interpreted as measure of uncertainty about the event at frequency f. In the field of nonlinear dynamics, approximate entropy [Hiborn, 94] is used to estimate the complexity of signals. A phase space (including embedding dimension, and delay time), however, is constructed before calculating approximate entropy, which is a difficult task. Second method is fractional spectral radius (FSR) [Broomhead et al., 1984], unfortunately the method needs to calculate the eigenvalues, which will spend too much time to get this indicator. In neurobiology, one often finds a correlation dimension of EEG signals, which appears to increase from sleep to waking states [e.g. Babloyantz, 1985]. Therefore the correlation dimension is a measure of complexity of EEG data. Unfortunately, the correlation dimension would be higher for a complete independent system, it is very clear that correlation dimension violates the criterion for complexity mentioned above. An additional well-known measure is Kolmogorov complexity [Kolmogorov, 1965]. This measure is appropriately low for completely regular strings, it is highest for random strings, and thus it also does not satisfy the above criterion for complexity as well [Tononi et al., 1994]. A simple method is to take AR model order [Sharma et al., 1997] as a measure of complexity of a time series, if a large number of coefficients are required to model, then the sequence may be called more complex than one which require only a few. The details will be discussed in the next section. Assumes that the value of the current sample, $y(t)$, in a data sequence, $y(1)$, $y(2)$, ..., $y(N)$, can be predicted as a linearly weighted sum of the p most sample values, $y(t-1)$, $y(t-2)$,..., $y(t-p)$, where p is the model order and is generally chosen to be much smaller than the sequence length, N. The predicted values of $y(t)$ can be expressed by $\tilde{y}(t) = -\sum_{i=1}^{p} a_i y(t-i)$, where weight a_i denote the ith coefficient of the pth-order model. The success of application of an AR model depends on the selection of an appropriate model order, p. Unfortunately, the correct model order for a given data sequence is not known in advance, thereby it is desirable to minimize the model's computational complexity by choosing the minimum value of p that adequately represents the signal. The finite sample information criterion (FSIC) [Broersen, 1998] is the best method for all AR order estimation and is independent of the choice of the maximum candidates order. In this chapter, FSIC is applied to estimate the order of EEG data. More detail of this method can be found in the reference [Broersen, 1998]. At the same time, a standard Levinson–Durbin recursion is used to estimate AR parameters.

Figure 2 shows two EEG signals with tonic-clonic seizures. This EEG contains control, pre-seizures, seizures and post-seizures. The two data can be found in [Quian *et al.*, 1997]. Epilepsy is a symptom of a brain disorder. The nerve cells in brain communicate with each other in a nice and calm way normally. However, an electrical thunderstorm strikes the brain, the nerve cells in brain are running out of control. An immense flood messages is sent out which may result uncontrolled speech, loss of consciousness, or convulsions of parts of your body, and so no. Because the brain uses electrochemical energy, any disruption of the electrical processes in the brain will cause abnormal functioning. Figure 3 shows the AR-order of EEG data over time that is post-processed by a joint probability [Pardey *et al.*, 1996]. It is found that the order of AR model decreases before seizure occurs (the stage of pre-seizure) at the seconds of 0-70 in Fig. 3(a) and at the seconds of 0-80 in Fig. 3(b); then, the order of AR model increases with the seizure procedure from Fig. 3(a). In Fig. 3(b), the order of AR model does not increase quickly, unlike Fig. 3(a). However, the order of AR model finally increases with seizure procedure. The characteristic of the post seizure in Fig. 3(b) is the same as one in Fig. 3(a), i.e. the order of AR model is very small at this stage. Therefore, the control, pre-seizures, seizures and post-seizures four different stages could be identified by using the order change of AR model in the light of these two case studies. It is worth noting that the complexity measure will fail when the time sequence under observation is random noise. In this case, model-order estimates will generally be low. This might be misinterpreted as a signal of "low complexity" even though, intuitively, the opposite definition applies to noise. Also, the measure relies on repetitive patterns found in the autocorrelation sequence. If such patterns do not exist, yet the sequence is not noise (e.g., a chaotic sequences); then the model-order-estimation procedure will fail. In such cases, again, the estimated model order is too small. Therefore, when we take the order of AR model as the complexity measure of an EEG signal, some possible failures should be considered.

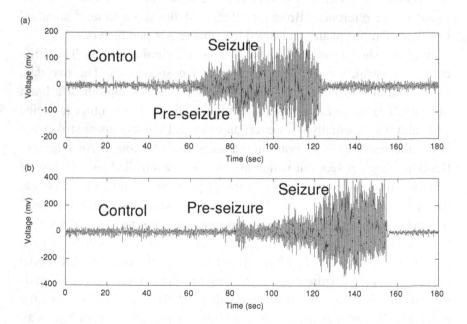

Figure 2: EEG signals with tonic clonic seizure preprocessed. (Data from [Quian *et al.*, 1997])

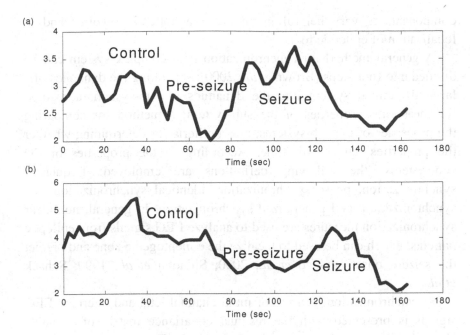

Figure 3: Orders of AR model post-processed.

4. Synchronization Measure

It is known that various levels of organization of brain tissue or between different parts of the brain exist synchronization. Synchronization phenomena have been recognized as a key feature for establishing the communication between different regions of the brain [Osvaldo *et al.*, 2001], and pathological synchronization as a main mechanism responsible of an epileptic seizure [Mormann *et al.*, 2000]. Some level of synchronization is normal neural activity, while too much synchrony may be pathological phenomena (such as epilepsy). Experimental models of epilepsy, as well as patient observations, suggest that the transition to ictal events is characterized by an abnormal increase in synchronization of neural activity. Thus, the changes in network synchrony may precede or follow the actual ictal event. Therefore, detection of synchronization and identification of causal relations between driving and response components is very helpful in anticipating epileptic seizures and in localization of epileptic foci.

A general method for synchronization of a complex system can be divided into four steps [Brown *et al.*, 2000]: separating the dynamics of a large dynamical system into the dynamics of sub-system; a method for measuring properties of the sub-system; a method for comparing the properties of the sub-systems; and a criteria for determining whether the properties agree with time. According to the properties of the sub-systems, the following definitions are employed: frequency/ synchronization, phase synchronization; identical synchronization; lag synchronization; and generalized synchronization. In general, nonlinear synchronization measures are used to analyze EEG signals from epileptic patients, which can be used to localize the epileptogenic zone and predict the seizure onset [Franco *et al.*, 2000; Shinobu *et al.*, 1999; Schack *et al.*].

A synchronization measure of multi-channel ictal and interictal EEG signals is presented with the residual covariance matrix of a multi-channel AR model [Franaszczuk *et al.*, 1999]. This measure considers the EEG signals in the framework of stochastic and deterministic models. Experimental models of epilepsy as well as patient observations suggest that the transition to ictal events is characterized by an abnormal increase in the synchronization of neural activity. Thus, the changes in network synchrony may precede or follow the actual ictal event. Such as, in [Iasemidis *et al.*, 1996] authors suggested that the onset of a seizure represents a spatiotemporal transition from a complex to a less complex (more ordered) state. The decrease of neuronal complexity in the primary epileptogenic area in preictal, ictal and postictal states is demonstrated [Lehnertz *et al.*, 1997]. An approach to EEG analysis has been motivated by the assumption that the EEG is generated by a stochastic system [Feuerstein *et al.*, 1992]. Multi-channel versions of the methods allow for the measurements of synchrony between the different spatial locations in brain using ordinary, partial [Gotman *et al.*, 1996]; or directed coherence [Wang *et al.*, 1992] and directed transfer function (DTF) [Franaszczuk *et al.*, 1999]. Measurements of these functions in preictal and ictal states show that the increased synchrony between the channels is close to the epileptogenic focus. The detail of the method is below. A pth-order vector AR process can be expressed as:

$$\mathbf{x}_t = \sum_{j=1}^{p} \mathbf{A}_j \mathbf{x}_{t-j} + \mathbf{e}_t \quad , \tag{3}$$

where $\mathbf{A}_j$ are $m \times m$ matrices of model coefficients, $\mathbf{x}_t$ is the vector of the multi-channel signal, and m is the number of channels, and $\mathbf{e}_t$ is the vector of multivariate zero mean uncorrelated white noise. The residual covariance $m \times m$ matrix $\mathbf{V_e}$ of noise vector $\mathbf{e}_t$ reflects the goodness of fit of a linear model to data. If the complexity of the system is large or the system is stochastic, the residuals will be larger. As a goodness of fit measure we are using $s = log(\det(\mathbf{V_e}))$, where the lower values of s reflects a better fit to a linear model, i.e. higher spatiotemporal synchronization. The quantity s can be interpreted as a synchronization measure of a system. If the channels are highly correlated, one channel can be predicted using other channels, the number of variables that describe dynamics of the system is lower, and the AR model is better fit, resulting in more small values of s. The minimum of s always occurs shortly after the onset of a seizure. This reflects the high regional synchrony between the channels and the very regular, almost periodic, rhythmic activity near the onset of a seizure. This result is in agreement with the reports of reduced complexity of brain electrical activity at the beginning of a seizure, and organized rhythmic activity (ORA) visible in the time-frequency analysis.

Herein, four sets collected from two healthy people and two patients with epilepsy (denoted A-D) are analyzed using multi-channel AR model of arbitrary order [Neumaier *et al.*, 2001], as shown in Fig. 4. The parameters and order of model are estimated by using a stepwise least squares algorithm, which is a computationally efficient method. Each set contains 100 single channel EEG segments of 23.6 sec duration. Sets A and B are relaxed in an awake state with eyes open and eyes close, respectively. Sets D is from within the epileptogenic zone; and sets C is from the hippocampal formation of the opposite hemisphere of the brain. Sets C and D contained the activity measured during seizure free intervals. The description of the data can be found in [Andrejak *et al.*, 2001].

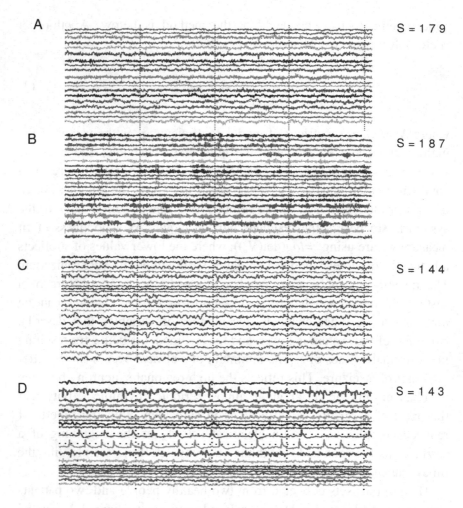

A   S = 1 7 9

B   S = 1 8 7

C   S = 1 4 4

D   S = 1 4 3

Figure 4: Part of EEG data of four sets. (Data from [Neumeier *et al.*, 2001]).

The *s* value of sets A-D that is calculated by using the above method are 179, 187, 144 and 143, respectively. The *s* value of the sets C and D is very close, and they are lower than the *s* value of the sets A and B. The lower values of *s* reflect a better fit to a linear model, i.e., higher spatiotemporal synchronization. The quantity *s* can be interpreted as a synchronization measure of the brain. The minimum of *s* always occurs shortly after the onset of a seizure, this reflects the high regional synchrony between the channels and the regular, almost periodic, rhythmic activity near the onset of a seizure for sets C and D. Sets A and B are from healthy people, the s value should be higher than the epilepsy patient's. The difference of sets A and B is caused by the different artifacts. From four cases, the synchronization measure with multi-channels AR mode of arbitrary order can possibly be applied to analyze the EEG signals for predicting epileptic seizures.

5. Conclusions

EEG signals have a wide variety of forms, such as the different effect to the signals because of the different ages or sex of the patients. Using traditional methods to extract the frequencies and amplitudes of the signals is not enough to represent the characteristics of system change.

Direct assessment of EEG signal complexity offers certain advantages in practical research. Complexity of EEG signals not only is an intuitive description, also makes the comparison between patients possible, as the complexity is insensitive to absolute measures (frequency or amplitude). In this chapter, we propose a simple method, the order of a time varying AR model, to measure the complexity of EEG signal. From two case studies, the complexity of EEG data can indicate the seizure approaching.

A multi-variable AR model is proposed to indicate the synchronization measure of multi-channels EEG signals. This synchronization measure of EEG data is based on the residual covariance matrix of a multi-channel autoregressive model, so as to present the characteristics of an EEG signal stochastically and deterministically.

The performances of these two methods are tested by some real cases; main conclusions can be summarized as follows:

The complexity measure with the AR model can show the change of an EEG signal; three stages of the tonic clonic seizure (pre-seizure, seizure and post-seizure) can be revealed by the complexity measure calculated with AR model order;

A synchronization measure of multi-channel EEG signal can be calculated by using the multi-variable AR model, the goodness of fit measure reflects the change of the epileptic seizure and normal condition.

References

Akay, M. (1994) Biomedical Signal Processing, Acad. Press, San Diego, 1994.

Anderson, C.W., Stolz, E.A., Sanyogita, S. (1998). Multivariate autoregressive models for classification of spontaneous electroencephalographic signals during mental tasks, *IEEE Trans. Biomedical Engineering*, 45, pp. 277-286.

Andrzejak, R.G., Lehnertz, K., Morman, F., Rieke, C., David, P. and Elger, C.E. (2001) Indications of nonlinear deterministic and finite-dimensional structures in time series of brain electrical activity: Dependence on recording region and brain state. *Phys. Rev. E*, 64, pp. 1-8.

Babloyantz, A., Salazar, J.M. and Nicolis, C. (1985). Evidence of chaotic dynamics of brain activity during the sleep cycle, *Phys. Lett. (A)*, 111, pp. 152–156.

Buzsaki, G. and Draguhn, A. (2004). Neuronal Oscillations in Cortical Networks, *Science*, 304, pp. 1926-1929.

Brown, R. and Kocarev, L. (2000). A unifying definition of synchronization for dynamical systems, *Chaos*, 10, pp. 344-349.

Broersen, P. (2000). Finite sample criteria for autoregressive order selection, *IEEE Trans. Signal Processing*, 48, pp. 3550-3558.

Broomhead, D. S. and King, G. P. (1986) Extracting qualitative dynamics from experimental data, *Physica D*, 20, pp. 217-236.

Carmeli, C, Knyazeva, M.G. Innocenti, G.M. De Feo, O. (2005). Assessment of EEG synchronization based on state-space analysis, *Neuroimage*, 25(2), pp. 339-54.

Engel AK, Fries P, Singer W. Dynamic predictions: oscillations and synchrony in top-down processing. Nat Rev Neurosci. 2001 Oct;2(10):704-16.

Feuerstein, G.C., Pham, D.T., Rondouin, G. (1992). On the tracking of rapid dynamic changes in seizure EEG, *IEEE Trans. Biomedical Engineering*, 39, pp. 952-958.

Ferrillo, F., Manolo, Beelke and Lino, Nobili. (2000). Sleep EEG synchronization mechanisms and activation of interictal epileptic spikes, *Clinical Neurophysiology*, 111, pp. 65-73.

Franaszczuk, P.J., Bergey, G.K. (1999). An autoregressive method for the measurement of synchronization of interictal and ictal EEG signal, *Biological Cybernetics*, 81, pp. 3-9.

Glass, L. (2001). Synchronization and rhythmic processes in physiology, *Nature*, 410, pp. 277-284.

Gotman, J. and Levtova, V. (1996). Amygdala-hippocampus relationships in temporal lobe seizures: a phase-coherence study, *Epilepsy Res.* 25, pp. 51-57.

Haykin, S. (1996). Adaptive Filter Theory. Prentice Hall, Englewood Cliffs, NJ.

Hiborn, R.C. (1994). Chaos and nonlinear dynamics, New York: Oxford Uni. Press.

Iasemidis, L.D. and Sackellares, J.C. (1996). Chaos theory and epilepsy, The Neuroscientist, 2, pp.18-126.

Jouny, C.C., Franaszczuk, P.J., Bergey, G.K. (2005). Signal complexity and synchrony of epileptic seizures: is there an identifiable preictal period? *Clin Neurophysiol.*, 116, pp. 552-558.

Kiymik, G.M.K, Akin, M, Alkan, A. (2001). AR spectral analysis of EEG signals by using maximum likelihood estimation, *Computers in Biology and Medicine*, 31, pp. 441-450.

Kong, X, Lou, X. and Thakor, N. (1997). Detection of EEG changes via a generalized Itakura distance, *Annual International Conference of the IEEE Engineering in Medicine and Biology - Proceedings*, 4, pp. 1540-1542.

Kolmogorov, A.N. (1965). Three approaches to the quantitative definition of information, *Inf. Trans*, 1, pp. 3–11.

Krystal, D., Prado, R. and West, M. (2000). New methods of time series analysis for non-stationary EEG data: Eigenstructure decompositions of time varying autoregressions, *Clinical Neurophysiology*, 110, pp. 1-10.

Lehnertz, K. and Elger, C.E. (1997). Neuronal complexity loss in temporal lobe epilepsy: effects of carbamazepine on the dynamics of the epileptogenic focus, *Electroenceph. Clin. Neurophysiol.*, 103, pp. 376-380.

Le Van Quyen, M. Soss, J. Navarro, V. Robertson, R. Chavez, M. Baulac, M. and Martinerie, J. (2005). Preictal state identification by synchronization changes in long-term intracranial EEG recordings, *Clin Neurophysiol.*, 116, pp.559-568.

Li, X., Guan, X. and Du, R. (2003). Using damping time for epileptic seizures detection in EEG, Modelling and Control in Biomedical Systems (Edited by Feng and Carson), Elsevier Ltd, pp. 255-258.

Li, X., Ouyang, G., Yao, X., and Guan, X. (2004). Dynamical characteristics of pre-epileptic seizures in rats with recurrence quantification analysis, *Physics Letters A* , 333, pp. 164–171.

Mormann, F. Lehnertz, K. David, P. Elger, C.E. (2000). Mean phase coherence as a measure for phase synchronization and its application to the EEG of epilepsy patients, *Physica D*, 144, pp. 358-369.

Neumaier, J. and Schneider, T. (2001). Estimation of parameters and eigenmodes of multivariate autoregressive models, *ACM Trans. Mathematical Software*, 27, pp. 27-57.

Osvaldo, A., Rosso, Susana, Blanco, Juliana, Yordanova, Vasil, Kolev, Alejandra, Figliola, Martin, Schürmann and Erol, B.(2001). Wavelet entropy: a new tool for analysis of short duration brain electrical signals, *Journal of Neuroscience Methods*, 105, pp. 65-75.

Pardey, J., Roberts, S. and Tarassenko, L. (1996). A review of parametric modelling techniques for EEG Analysis, *Med. Eng. Phys.* 18, pp. 2-11.

Patomäki, L., Kaipio, J.P. and Karjalainen, P.A. (1995). Tracking of nonstationary EEG with the roots of ARMA models, *Proc. 17 th Intl. Conf. IEEE EMBS.*

Quian, Q.R., Blanco, S., Rosso, O., Garcia, H. and Rabinowicz, A. (1997). Searching for hidden information with gabor transform in generalized tonic-clonic seizures, *Electroenceph. and Clin. Neurophysiol.*, 103, pp. 434-439.

Rissanen, J. (1986). A predictive least-squares principle, *IMA J. Math. Contr. Inform.* 3, pp. 211–222.

Rezek, A. and Roberts, S.J. (1998). Stochastic complexity measures for physiological signals analysis, *IEEE Trans. Biomedical Engineering*, 45, pp.1186-1191.

Sharma, A., Roy, R.J. (1997). Design of a recognition system to predict movement during anesthesia, *IEEE Trans. Biomedical Engineering*, 44, pp. 505-511.

Schack, A., Grieszbach, G. and Krause, G. (1999). The sensitivity of instantaneous coherence for considering elementary comparison processing. Part I: the relationship between mental activities and instantaneous EEG coherence, *International Journal of Psychophysiology*, 31, pp. 219-240.

Schloegl, A., Roberts, Pfurtscheller G. (2000). A criterion for adaptive autoregressive models, . World Congress on Medical Physics and Biomedical Engineering 2000.

Sharma, A. and Roy, R.J. (1997). Design of a recognition system to predict movement during anesthesia, *IEEE Trans. Biomedical Engineering*, 44, pp. 505-511.

Shinobu, K., Susumu, M., Kimiaki, U., Taeko, S. and Masako, K. (1999). Brainstem triggers absence seizures in human generalized epilepsy, *Brain Research*, 837, pp.277-288.

Steriade, M. and Timofeev, I. (2003). Neuronal plasticity in thalamocortical networks during sleep and waking oscillations, *Neuron.*, 37(4), pp. 563-76.

Stoica, P. and Moses, R. L. (1997). Introduction to Spectral Analysis, Prentice-Hall, Inc., New Jersey.

Thakor. N. V. and Tong, S. (2004). Advances in quantitative EEG analysis methods, Annual Review of Biomedical Engineering, 6, pp. 453-459.

Tseng, S., Chen, R. Chong, F. and Kuo, T. (1995). Evaluation of parametric methods in EEG signal analysis, *Medical Engineering & Physics*, 17, pp. 71-78.

Tononi, G., Sporns, O. and Edelman, G.M. (1994). A measure for brain complexity: relating functional segregation and integration in the nervous system, *Proc. Natl. Acad. Sci.* 91, pp. 5033–5037.

Tononi, G., Edelman, G.M. and Sporns, O. (1998). Complexity and coherency: integrating information in the brain, *Trends in Cognitive Sciences*, 2(12), pp. 474-484.

Traub, R.D., Spruston, N., Soltesz, I., Konnerth, A., Whittington, M.A. and Jefferys J.G.R. Gamma-frequency oscillations: a neuronal population phenomenon, regulated by synaptic and intrinsic cellular processes, and inducing synaptic plasticity, *Prog Neurobiol.*, 55(6), pp. 563-575.

Traub, R.D., Jefferys, J.G.R. and Whittington, M.A. (1999). Fast Oscillations in Cortical Circuits. MIT Press, Cambridge (MA).

Varela, F., Lachaux, J.P., Rodriguez, E., Martinerie, J. (2001). The brainweb: phase synchronization and large scale integration, *Nature Reviews Neuroscience*, 2, pp. 229-239.

Wang, X-J. (2003). Neural Oscillations: in Encyclopedia of Cognitive Science, MacMillan Reference Ltd , pp. 272-280.

Wang, G. and Takigawa, M. (1992). Directed coherence as a measure of interhemispheric correlation of EEG, *Int. J. Psychophysiol.*, 13, pp. 119-128.

Ward, G., Woodward, G., Stevens, A., and Stinson, C. (2003). Using overt rehearsals to explain word frequency effects in free recall, *Journal of Experimental Psychology: Learning, Memory, and Cognition*, 29(2), pp. 186-210.

CHAPTER 6

BAYESIAN FUSION OF SYNDROMIC SURVEILLANCE WITH SENSOR DATA FOR DISEASE OUTBREAK CLASSIFICATION

Jeffrey S. Lin[1], Howard S. Burkom[1], Sean P. Murphy[1], Yevgeniy Elbert[2], Shilpa Hakre[2], Steven M. Babin[1], Andrew B. Feldman[1]

[1]*The Johns Hopkins University Applied Physics Laboratory,*
[2]*Walter Reed Army Institute for Research*

Abstract

A novel Bayesian approach to fuse sensor data with syndromic sur-veillance data is presented for timely detection and classification of disease outbreaks, both natural and bioterror related. To validate the approach, we selected a natural disease, asthma, that is highly depend-ent on environmental factors, and for which we could identify transient periods of increased reactive airway attacks from medical diagnostic data. We treated such periods as outbreak surrogates and demonstrated that in the absence of historical training data, Bayesian belief networks with heuristically derived structures and conditional probability tables can provide statistically significant outbreak detection capability. We evaluated the benefit of the fusion of syndromic surveillance data with sensor data in two scenarios. The first was the augmentation of syn-dromic data with supporting, but nonspecific, sensor data. The second was the augmentation of a presumed highly specific sensor with syndromic data. Both approaches benefited from data fusion. This study demonstrates the potential of Bayesian fusion approaches to improve the detection and classification of bioterror attacks using syndromic data in conjunction with environmental and biowarfare sensor data.

1. Introduction

The threat of the use of biological agents in a terrorist attack has motivated the development of new technologies for both direct detection of an attack through biosensors and indirect detection through syndromic, pre-diagnostic indicators. To be effective against this threat, the developed systems must provide timely detection of an attack with sufficient sensitivity and specificity to enable deployment of effective and affordable surveillance programs. Advances in information management systems are now providing consumer data, such as over-the-counter pharmaceutical sales records, at daily and increasingly faster data transmission rates, thus further enabling the use of these data streams for the early detection of a disease outbreak caused by a biological weapon. These data, however, are noisy and nonspecific, and consequently, alerting algorithms derived using these data typically have high false alarm rates (low specificity) at the desired sensitivity. While improvements in biosensor technology are providing better tools for earlier detection of attacks at sites of concern, the coverage areas afforded protection are limited. On the other hand, the specificity of syndromic surveillance can be improved by using sensors monitoring the climate and the quality of air and water. For example, surges in respiratory pre-diagnostic indicators can be triggered by environmental factors as well as by infectious respiratory diseases. In this case, the supporting environmental data can be used to "explain away" a potential indication of a respiratory disease outbreak. The result can be a reduction in false alarms. In this chapter, we investigate the use of Bayesian fusion of these disparate data sources (syndromic [1] and environmental) to improve both the sensitivity and specificity of outbreak detection.

Syndromic surveillance is the monitoring of consumer and pre-diagnostic medical data to detect indications of disease outbreaks sufficiently early to permit life-saving intervention by the public health and medical communities. Drug purchases and medical visits categorized in syndrome groups of interest are counted and tracked for anomalous behavior. An important objective of syndromic surveillance is to identify the source of anomalies in the various data streams. This task is ideally performed by trained epidemiologists, however tools to help automate data fusion and event classification for epidemiologists are lacking. Ideally, the algorithms and models used to provide the event classifications would learn to recognize the event by training on real-world positive cases, with the trained classifier validated with real-world data. For disease outbreaks due to bioterrorism, there is a paucity of relevant data, and thus the development of classifiers for these events requires a more heuristic (knowledge, experience) based approach, in which the presumed data interrelationships are encoded directly using models. The necessary heuristics can be derived from medical and epidemiological expertise as well as the relevant literature.

In this study, we examined data fusion using a Bayesian Belief Network (BBN)-based model for detecting asthma "outbreaks". We use the term "outbreak" loosely here to refer to a sudden rise in reactive airway attacks in the population, even though such events are not truly outbreaks in the normal epidemiological sense. Specific asthma diagnosis counts provided a practical, verifiable example for studying the fusion of syndromic and environmental data, because it is well known that environmental factors such as high ozone levels and pollen counts can trigger asthma attacks [e.g. 2–5]. In addition, specific, timely diagnosis data (International Classification of Diseases-9 [ICD-9] codes) were available to provide the often elusive "ground truth" against which predictions of the BBN can be evaluated.

This study used daily syndromic surveillance data from the Electronic Surveillance System for the Early Notification of Community-Based Epidemics (ESSENCE)[6], developed by the Walter Reed Army Institute for Research and extended in collaboration with The Johns Hopkins University Applied Physics Laboratory. ESSENCE II, a regional version of the ESSENCE systems, provides syndromic surveillance for the U.S. National Capital Region, and includes data from the Baltimore, MD – Washington D.C. metropolitan area. The ESSENCE data were acquired from both military and civilian facilities. We obtained the air quality data from the Environmental Protection Agency, with the air quality indices (AQI) provided for each county/municipality. The allergen levels, measured by the US Army Centralized Allergen Extract Laboratory at the Walter Reed Army Medical Center in Washington, DC, were available for only one location in the monitored area.

The primary objectives of this study were: (1) to demonstrate the feasibility of using heuristically derived BBNs to perform data fusion for detecting disease outbreaks in the absence of training data; and (2) to investigate the potential for improvement of detection performance through fusion of sensor data with syndromic data relative to the use of either data type alone.

2. Approach

2.1. *Bayesian belief networks*

The approach we chose to fuse syndromic and sensor data to detect and classify disease outbreaks was driven by the inherent uncertainties in the underlying data. Syndromic data are noisy, and reflect the influence of innumerable factors that affect, e.g., the purchasing behavior of over-the-counter and prescription medications and the decision when to seek medical care. Environmental data were only acquired at specific locations, and thus the readings may or may not reflect the levels of pollutants or allergens in other nearby areas, depending on local weather patterns. In addition, although data was recorded at regular intervals, prolonged intervals of missing data were common. The need to robustly fuse these uncertain evidentiary data and to accommodate instances in which some evidence may be missing suggested the use of model-based Bayesian Belief Networks (BBNs).

The BBN approach provides a compact encoding of a joint probability distribution. A BBN is typically visualized using a directed acyclic graph, with nodes representing random variables, and directed edges (or lack of them) representing probabilistic dependence (or independence). For a discrete-valued BBN, each node is associated with a conditional probability table (CPT) describing the conditional dependence of the value of the node on the values of its parent nodes. Figure 1 shows the structure of the BBN we developed to detect asthma "outbreaks" from syndromic and/or environmental sensor data. The value of the *Asthma* node at the top of Fig. 1 is the inferred probability of an ongoing or impending asthma outbreak based on the values of the

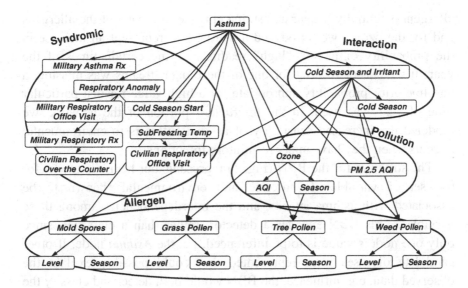

Figure 1: BBN structure for asthma detection through fusion of syndromic and environmental data.

evidence variables. Nodes that are not the source of any directed edges—an example is Ozone Air Quality Index (AQI) in the figure—are the variables whose values are assigned as evidence. We have used the Bayes Net Toolbox for Matlab [7] for our prototype implementation.

The BBN structure in Fig. 1 shows several classes of evidence fused by the network. The *Syndromic* group included evidence of anomalies in syndromic data streams, and a node representing the influence of the cold/flu season, which can be a trigger for asthma attacks. In the *Pollution* portion of the network, we restricted the data to include only the AQI of ozone and of particulate matter ≤ 2.5 μm (PM2.5) in aerodynamic diameter. We omitted carbon monoxide and sulfur dioxide measurements because these rarely reach unhealthy levels in the National Capital Region. The *Allergen* portion of the network included concentrations of mold spores and three types of pollen. For the effect of cold weather alone, the network has an evidence node representing a binary state describing the maximum temperature for the day: sub-freezing or above freezing. A subfreezing temperature can, like an allergen, potentially trigger an asthma episode. For each of the allergens and for the ozone, we introduced a heuristic to represent an increase in the probability of having "high" readings during certain times of the year. If the sensor reading for one of these data streams was missing, a baseline prior probability appropriate for a high value for that particular time of year was used. In the *Interaction* portion of the network, we modeled the increased sensitivity of asthma sufferers to environmental triggers when they have upper respiratory infections.

The structure for the BBN (Fig. 1) represents our heuristic model for the set of variables (syndromic and environmental) known to be associated with asthma attacks, and the interdependencies among these variables. This BBN encodes a detector, rather than a classifier, since only one node's value is to be inferenced, i.e. the *Asthma* node. If other candidate diseases were represented that could possibly explain the observed data, e.g. influenza, the BBN would both detect and classify the diseases. Not represented in Fig. 1 are the CPTs associated with each node. The conditional probabilities in these tables could be estimated using data. The quality of such estimates depends on both the stationarity (stability of underlying statistical processes) and the quantity (sample size for parameter estimation) of the data. The model of a nonstationary process can be improved only by understanding the causes of the nonstationarity and modeling those effects.

If the data set is extremely limited, it is possible to "bootstrap" a BBN with heuristic CPTs, derived from initial beliefs and domain expertise, and then updated appropriately as the data becomes available. Table 1 lists four representative CPTs for our heuristic network for the National Capital Region. In the CPT for the *Respiratory Anomaly* node, Table 1(a), we estimated that 70% of surges in asthma cases result in a detectable anomaly in prediagnostic data streams that we categorized in the respiratory syndrome. This estimate accounts for factors such as asthma sufferers maintaining a supply of appropriate medications. To complete the table, we estimated that 3% of the time that there is not an ongoing asthma outbreak, there is an anomaly in respiratory syndrome data streams due to other causes. This value is a function of the sensitivity of the anomaly detectors and roughly corresponds to a

Table 1. Representative CPTs for BBN nodes for asthma detection through fusion of syndromic and environmental data.

a)

Asthma	P(Respiratory Anomaly)	
	True	False
True	0.70	0.30
False	0.03	0.97

c)

Asthma	Cold Season and Irritant	P(PM 2.5)	
		True	False
True	True	0.06	0.94
True	False	0.06	0.94
False	True	0.0067	0.9933
False	False	0.0133	0.9867

b)

Asthma	P(Cold Season and Irritant)	
	True	False
True	0.40	0.60
False	0.05	0.95

d)

Asthma	Cold Season and Irritant	P(Tree Pollen)	
		True	False
True	True	0.05	0.95
True	False	0.10	0.90
False	True	0.0067	0.9933
False	False	0.0133	0.9867

background anomaly alert rate of 1 per month. In Table 1(b) for the *Cold Season and Irritant* node, we estimated that 40% of asthma outbreaks were caused by the combination of respiratory infections and environmental triggers, and 5% of the time there is an environmental trigger during the cold season that does not coincide with an asthma outbreak. The *PM 2.5* CPT, Table 1(c), is more complex given that it has two parent nodes, *Asthma* and *Cold Season and Irritant*, yielding four combinations of the two binary variables. We derived the table entries by estimating that 12% of all asthma outbreaks are due to high PM 2.5 pollution. We further assumed that half of these (6%) occur during the cold/flu season and half (6%) during the remainder of the year. In the lower half of the table where *Asthma* is false, we estimated 6 days per year with high PM 2.5 pollution that does not cause a detectable asthma outbreak. These days are evenly distributed throughout the year, assuming a 4 month cold/flu season. Similar reasoning was used to generate the *Tree Pollen* CPT in Table 1(d).

If the amount of data available is extensive, heuristic CPTs are presumably not required and maximum likelihood estimates (MLE) of conditional probabilities obtained from the data should be accurate. To compute MLEs for our network, an expectation maximization (EM) step was required because there were missing data in some instances. The iterative EM step typically converges to a local optimum, so CPTs calculated with MLE using this step depend on the initial (seed) conditional probability values used.

The computational burden for performing inferences using BBNs increases exponentially with the size of the network. We forestall scaling problems with a top-down approach, using aggregated data and modeling high-level relationships between data streams, to keep the size of the network comparatively small. Our non-optimized Matlab implementation of the asthma BBN required a fraction of a second on a desktop computer to inference the value of the *Asthma* node from each day's data. All estimations of the CPTs from the data are performed off-line and are less time/resource constrained. As the resolution of the model increases, however, computational limits eventually become significant. At the other extreme, for example, bottom-up approaches to disease surveillance model the disease state of each individual in a population of millions. Equivalence class and incremental updating methods are used to reduce the computational burden [8]. Additionally, approximate methods offer the ability to greatly expand the usable BBN complexity with the perhaps minimal sacrifice of an exact solution [9].

2.2. *Syndromic data*

The syndromic data records collected by the ESSENCE systems were tagged with many fields, such as zip code, a de-identified health provider code, and age and sex where available. (The ESSENCE systems do not have access to personal identifying information.) For this initial analysis, we aggregated all of the syndromic data across the entire monitored area of the National Capital Region, without stratifying by age, sex, or subregion. The daily data counts used as "ground truth" for the evaluation of the BBN detector were derived from all of the ESSENCE military and civilian office visit records with ICD-9 codes specific for asthma. The raw counts of these asthma-specific records were processed using a hypothesis test developed for ESSENCE to identify statistical anomalies. For the asthma diagnosis count data, the test statistic was the standardized residual of a multivariate regression model with a sliding baseline. The covariates used for this regression were: the number of providers represented in each day's data records; an indicator variable for weekends or holidays; an indicator variable for the day immediately after a weekend or holiday; and the linear trend in the baseline. The standardized residuals were assumed to form a zero-mean Gaussian distribution. For each day, a test of the null hypothesis that the current standardized residual fits this distribution was applied and a p-value assigned to the day. We identified a statistical anomaly if this null hypothesis was rejected. For the purposes of this study, a ground-truth outbreak was formally defined as the interval beginning 3 days prior to the asthma-ICD-9 anomaly day and ending the day after the anomaly. Using this definition, we found 17 distinct asthma outbreaks in the two years (2002 and 2003) of data used for the study.

Using the above methodology, the increases in asthma diagnoses that were treated as outbreaks were automatically identified from daily counts of ICD-9 codes, without human interpretation or application of heuristics. This algorithmic approach eliminated the possibility of skewing our results by inadvertently using data included as evidence in the BBN to define the event, such as coincidence of an asthma anomaly with the start of cold/flu season or high pollen times of year. We anticipated that some of these outbreaks should not be detected by our BBN because we did not include data encompassing all possible triggers for asthma.

The syndromic datasets used as evidence for the BBN detector were aggregated over the entire region without age, gender or other stratification, similar to the aggregation of the ground truth data. Five groups of syndromic data were selected from ESSENCE: civilian over-the-counter drug sales; the analogous military general prescription drug (Rx) sales; civilian and military office visits; and military asthma-specific prescription sales. For the more general syndromic groups, counts were filtered to include only visits or sales classified in a general respiratory category. The *Military Asthma Rx* group included only non-refill prescriptions. This data stream was delayed by 1 day relative to the asthma-specific diagnostic counts used as ground truth. By using the previous day's asthma-specific prescriptions sales, we were not trivially detecting an increase in asthma diagnoses by inspecting prescription sales potentially directly resulting from those diagnoses. The raw data counts for the above syndrome groups were processed with the same multivariate linear regression used to identify the asthma outbreaks. To improve the robustness of the BBN performance with respect to the choice of the threshold selected to identify an anomaly in a syndrome group, an additional processing step was performed to map the p-value derived from the hypothesis test to a probability that the p-value is anomalous. We used a sigmoidal transformation assigning a p-value of 0.95 to an anomaly probability of 0.50, and a p-value of 0.99 to an anomaly probability of 0.90. The distinction between these probabilities is subtle, but important. This transformation encoded a heuristic that a residual exceeding 95% of the expected residuals would be a "true" anomaly 50% of the time, and a residual exceeding 99% of the expected residuals would be a "true" anomaly 90% of the time. In contrast, using a hard threshold of p-value of 0.99 would map residual probabilities 0.99 to zero probability of being an anomaly, and >0.99 to a probability of one for being an anomaly. Our formulation reduces the sensitivity of results to the particular choice of the threshold p-value. The continuous value of probability of an anomaly for the syndromic evidence was assigned to the discrete-valued nodes (anomaly, no anomaly) as soft evidence in the Bayes Net Toolbox.

2.3. Environmental data

All environmental data were aggregated over the entire monitored region similar to the syndromic data. The allergen data, recorded at only one location in Washington D.C., did not require any aggregation. The AQI data, however, came from several counties. We selected a simple aggregation algorithm for these data in which the maximum AQI value over the entire region was chosen for each pollutant. For simplicity, we defined our environmental variables in the BBN to be binary valued, high or low. Allergen level measurements greater than moderate

(National Allergy Bureau Scale [10], Table 2) were assigned a value of high. AQI values greater than 150 were likewise assigned a value of high (Table 3).

Based on our review of the literature on the environmental triggers of asthma [2–5, 11–56], we assigned expected delays between high levels of a pollutant or allergen and a detectable surge in asthma diagnoses. An increased probability of an asthma outbreak was expected 3 days after a high ozone measurement, and 1 day after high pollen or mold levels. Thus, in presenting the data to the BBN, we shifted the ozone, pollen, and mold data by the appropriate number of days.

Table 2. National Allergy Bureau scale for counts per cubic meter of allergens. [8]

Level	Mold	Grass pollen	Tree pollen	Weed pollen
Absent	0	0	0	0
Low	1 - 6499	1 - 4	1 - 14	1 - 9
Moderate	6500-12999	5 - 19	15 - 89	10 - 49
High	13000 - 49999	20 - 199	90 - 1499	50 - 499
Very high	≥ 50000	≥ 200	≥ 1500	≥ 500

Table 3. US EPA Air Quality Index (AQI) Color Scale.

Index values	Color level	Descriptor
1–50	Green	Good
51–100	Yellow	Moderate
101–150	Orange	Unhealthy for Sensitive Groups
151–200	Red	Unhealthy
201–300	Purple	Very Unhealthy
301–500	Maroon	Hazardous

2.4. Test scenarios

In our first scenario, we examined whether the fusion of environmental sensor data with non-asthma-specific syndromic data could improve outbreak detection performance over that using the syndromic data alone. This scenario investigated the applicability of the BBN when environmental data was used to provide alternative explanations for the observed syndromic data. By analogy, it is also relevant to the case where non-specific, spatially distributed, bioweapon agent sensors with high false-alarm rates are used to provide distributed high sensitivity to a possible bioterror attack.

In our second scenario, we evaluated the use of syndromic data to confirm the beginning of an outbreak due to an environmental trigger that was detected by sensors. In this scenario, the sensors used typically would have low false alarm rates, but would be highly specific to the agent triggering the outbreak (although these conditions did not apply to the sensor data in this study). Such a system would be useful when actual human infection is not guaranteed when the threat agent is detected, and an effective and affordable response strategy requires threat confirmation and estimates of outbreak spatial distribution and extent.

2.5. Evaluation metrics

We evaluated the performance of our BBN detector for asthma using receiver operating characteristic (ROC) curves [57]. The choice of the decision threshold on the posterior probability for the *Asthma* node affects both the probability of detection and the probability of false alarm. ROC curves summarize the trade-off between these metrics as this decision threshold value is varied. The operational decision threshold is selected as the point on the ROC curve with the desired system performance. For this analysis, ROC curves were plotted parametrically as the probability of detection against the probability of false alarm, with the decision threshold serving as the parameter.

Our ROC analysis was supplemented by an additional significance statistic, which represents the probability of obtaining equivalent or better performance simply by chance. The significance of BBN detector performance at each operating point on the ROC curve was estimated using two Monte Carlo approaches. The first approach randomized the dates of the ground-truth outbreaks used to evaluate success. This process destroyed all correlations between the input data and the ground-truth outbreaks. The second approach randomized the CPT values in the BBN. Because only CPTs were randomized while the BBN structure was preserved, this approach measured the significance of the utilized CPTs for our chosen BBN structure. ROC curves were generated for each of 10,000 iterations for both approaches. The significance value at each operating point was defined by counting the number of ROC curves from the 10,000 that have a better false alarm rate for a detection probability at least as high as the operating point detection probability.

An important metric for demonstrating the value of any syndromic surveillance technique is the promptness of the detection. Earlier detection can potentially save lives through earlier and more effective response to the disease outbreak. In the absence of syndromic surveillance, medical diagnosis of individuals with the disease serves as the indicator of an outbreak. We, therefore, calculated the promptness by measuring the difference in days between an asthma detection date using the BBN and date of the detected the anomaly in asthma-specific ICD-9 data streams. Because successful detections were defined as between 3 days before and 1 day after the detected ICD-9 anomaly, these are the limits of the possible promptness values.

3. Results

3.1. *Scenario 1*

We first tested the validity and utility of a network derived entirely from heuristics by developing a BBN using whatever knowledge we could obtain from the literature [2–5, 10–55] and our in-house medical and epidemiological expertise. The ROC curve for this BBN, with heuristically derived CPTs, evaluated for 2002 (the first of the two years of syndromic and environmental data) is shown in Fig. 2(a). A manageable false alarm interval for public health epidemiologists is about once every 30 days. The false alarm rate corresponding to this interval is indicated on the abscissa in Fig. 2(a). At this false alarm rate, the heuristic BBN detected 3 of the 10 ground-truth asthma outbreaks. The significance of the results relative to random CPTs and random outbreaks at this operating point was ~0.03 [Fig. 2(b)]. The ROC curve for the BBN after the CPTs have been MLE trained on the first year of data from the heuristic initial state shows that the trained network detected 3 more outbreaks at about the same false alarm rate [Fig. 2(a)].

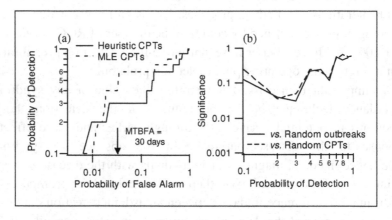

Figure 2: (a) ROC curves for asthma detection in year 1. BBN with heuristic CPTs and MLE trained CPTs. (b) Significance of ROC curve for BBN with heuristic CTPs in year 1.

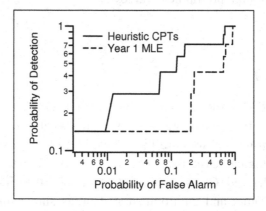

Figure 3: ROC curves for asthma detection in year 2. BBN with heuristic CPTs and CPTs MLE trained with year 1 data.

Table 4. Meteorological data from 2002 and 2003 at Reagan National Airport in Washington, D.C. [58]

Yearly statistic	2002	2003
Total rainfall	34.33 in.	59.53 in.
Total snowfall	10.3 in.	39.5 in.
Mean daily maximum temperature	68.1 °F	64.1 °F
Highest daily maximum temperature	100 °F	94 °F
Number of days with temperature $\leq 32°F$	2	11
Number of days with temperature $\geq 90°F$	56	20

The ROC curves for these same networks tested against the second year of data indicate that the performance of the MLE trained network was substantially worse than that of the heuristic BBN (Fig. 3). These results suggest that the characteristics of, or relationship between, the data streams changed significantly between the two years. This observation is supported by meteorological data that show the year 2002 was hot and dry relative to 2003 (Table 4). Indeed, the area was experiencing a severe drought in 2002, which ended in 2003. The fact that the heuristic BBN performed more consistently across the two years than a trained network indicates that the understanding on which this BBN was designed provides a more robust model than with probabilities learned from a limited training set. In a combined analysis of the two years, the heuristic BBN detected 6 of the 17 ground-truttbreaks at a false alarm rate of 1 every 40 days (Fig. 4). The significance values of the results relative to random CPTs and randomh ou outbreaks at this operating point were < 0.01. The benefit of fusing the sensor information (Pollution and Allergens in Fig. 1) with the syndromic data was demonstrated by the reduced performance of the ROC curve when the sensor data were removed [Fig. 4(a)].

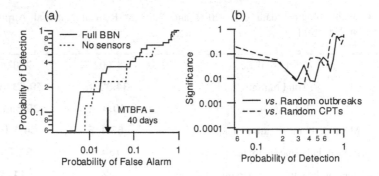

Figure 4: (a) Comparison of ROC curves for asthma detection BBNs with heuristic CPTs fusing both syndromic and sensor data (Full BBN) and syndromic data alone (No sensors) for years 1 and 2. (b) Significance of heuristic Full BBN ROC curve.

3.2. *Scenario 2*

In the second scenario, we assumed the sensor data to be highly-specific with very few false alarms. We thus queried the output of the BBN only when a sensor indicated a detection. In our case, we considered a sensor detection to be any "high" environmental sensor reading as defined previously. Of the original 17 ground-truth outbreaks, 9 were concurrent with sensor detections. If during one of these 9 outbreaks the BBN *Asthma* node posterior probability exceeded the decision threshold during a ground-truth asthma outbreak, we succeeded in confirming the sensor reading. We falsely confirmed a high sensor reading if there was not a concurrent ground-truth asthma outbreak. The false alarm rate has a different meaning in the ROC curve for scenario 2 compared to the previous ROC analysis [Fig. 5(a)]. Because the syndromic data were being used to confirm sensor detections, the false alarm rate is the average number of false confirmations per sensor detection, as compared to the average number of false detections per day. At the operating point of detecting 5 out of 9 outbreaks (56%) and a false confirmation every ~8 sensor detections, 8% of BBNs with random CPTs performed better, and < 1% of randomly assigned outbreaks yielded better results [Fig. 5(b)].

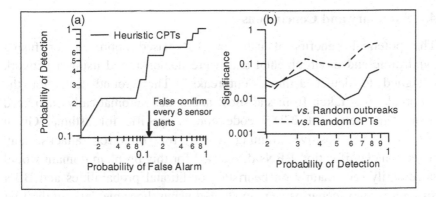

Figure 5: (a) Comparison of ROC curves for asthma detection BBNs with heuristic CPTs fusing both syndromic and sensor data (Full BBN) and syndromic data alone (No sensors) for years 1 and 2. (b) Significance of heuristic BBN ROC curve.

3.3. *Promptness*

The promptness of the BBN detections plotted as a histogram indicated that in both scenarios, roughly 70% of the BBN detections provided an early warning relative to the asthma-specific ICD-9 data streams (Fig. 6).

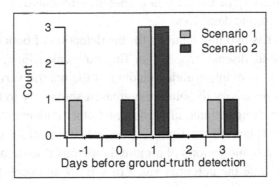

Figure 6: Histograms of the promptness with which the BBNs detected an asthma outbreak for the two test scenarios.

4. Summary and Conclusions

The potential benefits of using a BBN-based approach for fusing syndromic and environmental data were demonstrated using a network designed to detect asthma "outbreaks". The ground truth for the outbreaks was taken from statistical anomalies automatically calculated using time series of ICD-9 code counts specific for asthma. Given the lack of positive training cases for bioterror attacks, any detection/classification BBNs developed for this problem domain would necessarily rely mainly on heuristic conditional probabilities and BBN structure. Two scenarios were evaluated using the same data. In the first scenario, syndromic and sensor data were fused using an entirely heuristically derived BBN to detect 7 of the 17 ground-truth asthma outbreaks with a less than once a month false alarm rate. Training of BBN parameter values based on one year of data improved performance for that one year, but proved not to be as robust as the heuristic values when applied to data in the subsequent year. This suggests that heuristics derived from epidemiological and medical knowledge may be the better option when there is a paucity of training data. In the second scenario, syndromic data were used to confirm sensor detections of an asthma trigger. An all-heuristic BBN confirmed 56% of the sensor detections, with a false confirmation rate of roughly 13%. Finally, the power of fusing environmental data with syndromic data was demonstrated by the improved sensitivity of the asthma outbreak BBN classifier compared to one using syndromic data alone.

This study has two implications for the detection of both bioterror and emerging natural disease (e.g. avian flu) outbreaks. First, in detecting such outbreaks, a complete understanding of the natural variability of the background is necessary to obtain the greatest sensitivity to the outbreak. Models for detecting asthma, influenza, and other endemic "background" diseases can explain fluctuations in the data that might otherwise be mistaken for early indications of a more dangerous disease outbreak. Our results demonstrate the potential value of a BBN approach for modeling these background diseases.

The second implication is that our data fusion approach can be used to confirm an outbreak resulting from an environmental trigger prior to medical diagnosis. While we demonstrated this using asthma as a surrogate (however flawed), for the more dangerous disease outbreaks, we are mindful that its applicability requires further investigation and elucidation. Significantly, the favorable results we obtained in our studies were based on BBNs derived entirely from heuristic knowledge. The efficacy of heuristic knowledge in the absence of training data, and the fusion of highly aggregated data streams, suggest that a BBN approach for disease surveillance can be readily expanded to other diseases using solely epidemiological and medical expertise and literature, while remaining computationally tractable for large numbers of background and targeted diseases.

Acknowledgements

The authors thank Dr. I-Jeng Wang and Mr. Anshu Saksena for their valuable insights and expertise on Bayesian decision theory and Ms. Erin Symonds for her contributions to the data mining efforts. This work was funded by the Joint Program Executive Office for Chemical and Biological Defense of the U.S. DoD, and is based heavily on the results of concurrent research and development performed on JHU/APL research funds.

References

[1] H. S. Burkom, Y. Elbert, A. Feldman, and J. Lin, "Role of Data Aggregation in Biosurveillance Detection Strategies with Applications from the Electronic Surveillance System for the Early Notification of Community-Based Epidemics," MMWR, Vol 53, No SU01; pp. 67-73, 2004.

[2] W. Anderson, G. J. Prescott, S. Packham, J. Mullins, M. Brookes, and A. Seaton, "Asthma admissions and thunderstorms: a study of pollen, fungal spores, rainfall, and ozone," Q J Med, Vol 94, pp. 429-433, 2001.

[3] P. E. Tolbert, J. A. Mulholland, D. L. Macintosh, F. Xu, D. Daniels, O. J. Devine, B. P. Carlin, M. Klein, J. Dorley, A. J. Butler, D. F. Nordenberg, H. Frumkin, P. B. Ryan, and M. C. White, "Air quality and pediatric emergency room visits for asthma in Atlanta, Georgia," American J. Epidemiol., Vol. 151(8), 798-810, 2000.

[4] J. F. Gent, E. W. Triche, T. R. Holford, K. Belanger, M. B. Bracken, W. S. Beckett, and B. P. Leaderer, "Association of low-level ozone and fine particles with respiratory symptoms in children with asthma," JAMA, Vol. 290, pp. 1859-1867, 2003.

[5] D. H. Jaffe, M. E. Singer, and A. A. Rimm, "Air pollution and emergency department visits for asthma among Ohio Medicaid recipients, 1991-1996," Environmental Research, Vol. 91, 21-28, 2003.

[6] J. S. Lombardo, J. A. Pavlin, *et al.* "A systems overview of the Electronic Surveillance System for the Early Notification of Community-Based Epidemics (ESSENCE II)," J Urban Health 2003, 80(2 Suppl 1), pp. 32–42.

[7] K. Murphy, http://www.ai.mit.edu/~murphyk/Software/BNT/bnt.html

[8] G. F. Cooper, D. H. Dash, J. D. Levander, W-K Wong, W. R. Hogan, M. M. Wagner, "Bayesian Biosurveillance of Disease Outbreaks," Proceedings of the 20th Annual Conference on Uncertainty in Artificial Intelligence (UAI-04), pp. 94-103, 2004.

[9] R. E. Rosales, T. S. Jaakkola, "Focused Inference," Proceedings of AI & Statistics 2005, http://www.gatsby.ucl.ac.uk/aistats/fullpapers/164.pdf.

[10] http://www.aaaai.org/nab/index.cfm?p=reading_charts

[11] H. A. Burge, "An update on pollen and fungal spore aerobiology," J. Allergy Clin Immunol., Vol. 110, pp. 544-552, 2002.

[12] R. T. Burnett, R. E. Dales, M. E. Raizenne, D. Krewski, P. W. Summers, G. R. Roberts, M. Raad-Young, T. Dann, and J. Brook, "Effects of low ambient levels of ozone and sulfates on the frequency of respiratory admissions to Ontario hospitals," Environmental Research, Vol. 65, pp. 172-194, 1994.

[13] N. Chuersuwan, B. J. Turpin, and C. Pietarinen, "Evaluation of time-resolved PM2.5 data in urban/suburban areas of New Jersey," J. Air Waste Manag Assoc, Vol 50(10), pp. 1780-1789, 2000.

[14] H. Desqueyroux, J.-C. Pujet, M. Prosper, F. Squinazi, and I. Momas, "Short-term effects of low-level air pollution on respiratory health of adults suffering from moderate to severe asthma," Environmental Research, Section A, Vol 89, 29-37, 2002.

[15] R. J. Delfino, H. Gong, Jr, W. S. Linn, E. D. Pellizzari, and Y. Hu, "Asthma symptoms in Hispanic children and daily ambient exposures to toxic and criteria air pollutants," Environ Health Perspect., Vol 111, pp. 647-656, 2003.

[16] A. M. Donoghue, and M. Thomas, "Point source sulphur dioxide peaks and hospital presentations for asthma," Occup. Environ. Med., Vol. 56, pp. 232-236, 1999.

[17] D. M. Fleming, K. W. Cross, R. Sunderland, and A. M. Ross, "Comparison of the seasonal patterns of asthma identified in general practitioner episodes, hospital admissions, and deaths," Thorax, Vol. 55, pp. 662-665, 2000.

[18] V. R. Fuchs, and S. R. Frank, "Air pollution and medical care use by older Americans: a cross-area analysis," Health Affairs, Vol. 21(6), 207-214, 2002.

[19] W. E. Berger, "Overview of allergic rhinitis," Ann. Allergy Asthma Immunol., Vol. 90(Suppl 3), pp. 7-12, 2003.

[20] S. Hajat, H. R. Anderson, R. W. Atkinson, and A. Haines, "Effects of air pollution on general practitioner consultations for upper respiratory diseases in London," Occup. Environ. Med., Vol. 59, pp. 294-299, 2002.

[21] S. Hajat, A. Haines, R. W. Atkinson, S. A. Bremner, H. R. Anderson, and J. Emberlin, "Association between air pollution and daily consultations with general practitioners for allergic rhinitis in London, United Kingdom," American J. Epidemiol., Vol. 153(7), pp. 704-714, 2001.

[22] S. K. Heath, J. Q. Koenig, M. S. Morgan, H. Checkoway, Q. S. Hanley, and V. Rebolledo, "Effects of sulfur dioxide exposure on African-American and Caucasian asthmatics," Environmental Research, Vol. 66, pp. 1-11, 1994.

[23] D. Howel, R. Darnell, and T. Pless-Mulloli, "Children's respiratory health and daily particulate levels in 10 nonurban communities," Environmental Research, Section A, Vol 87, pp. 1-9, 2001.

[24] J.-S. Hwang, and C.-C. Chan, "Effects of air pollution on daily clinic visits for lower respiratory tract illness," American J. Epidemiol., Vol. 155(1), pp. 1-10, 2002.

[25] H. R. Anderson, S. A. Bremner, R. W. Atkinson, R. M. Harrison, and S. Walters, "Particulate matter and daily mortality and hospital admissions in the west midlands conurbation of the United Kingdom: associations with fine and coarse particles, black smoke and sulphate." Occup. Environ. Med., Vol. 58, pp. 504-510, 2001.

[26] B. B. Jalaludin, B. I. O'Toole, and S. R. Leeder, "Acute effects of urban ambient air pollution on respiratory symptoms, asthma medication use, and doctor visits for asthma in a cohort of Australian children," Environmental Research, Vol. 95, pp. 32-42, 2004.

[27] A. B. Kay, "Allergy and allergic diseases," New Engl. J. Med., Vol. 344(1), pp. 30-37, 2001.

[28] D. Kimes, E. Levine, S. Timmins, S. R. Weiss, M. E. Bollinger, and C. Blaisdell, "Temporal dynamics of emergency department and hospital admissions of pediatric asthmatics," Environmental Research, Vol 94, pp. 7-17, 2004.

[29] J. Q. Koenig, K. Jansen, T. F. Mar, T. Lumley, J. Kaufman, C. A. Trenga, J. Sullivan, L.-J.S. Liu, G. G. Shapiro, and T. V. Larson, "Measurement of offline exhaled nitric oxide in a study of community exposure to air pollution," Environ Health Perspect., Vol 111, pp. 1625-1629, 2003.

[30] M. Lin, Y. Chen, R. T. Burnett, P. J. Villenueve, and D. Krewski, "Effect of short-term exposure to gaseous pollution on asthma hospitalisation in children: a bidirectional case-crossover analysis," J. Epidemiol. Community Health, Vol. 57, pp. 50-55, 2003.

[31] G. Norris, S. N. YoungPong, J. Q. Koenig, T. V. Larson, L. Sheppard, and J. W. Stout, "An association between fine particles and asthma emergency department visits for children in Seattle." Environmental Health Perspectives, Vol. 107(6), pp. 489-493, 1999.

[32] G. E. Packe, and J. G. Ayres, "Asthma outbreak during a thunderstorm," The Lancet, Vol. 8448(2), pp. 199-204, 1985.

[33] J. L. Peacock, P. Symonds, P. Jackson, S. A. Bremner, J. F. Scarlett, D. P. Strachan, and H. R. Anderson, "Acute effects of winter air pollution on respiratory function of schoolchildren in southern England," Occup. Environ. Med., Vol. 60, pp. 82-89, 2003.

[34] A. Pitard, A. Zeghnoun, A. Courseaux, J. Lamberty, V. Delmas, J. L. Fossard, and H. Villet, "Short-term associations between air pollution and respiratory drug sales," Environmental Research, Vol 95, pp. 43-52, 2004.

[35] A. Ponka, "Asthma and low level air pollution in Helsinki," Arch. Environ. Health, Vol 46(5), pp. 262-270, 1991.

[36] F. Sartor, R. Snacken, C. Demuth, and D. Walckiers, "Temperature, ambient ozone levels, and mortality during Summer, 1994, in Belgium," Environmental Research, Vol. 70, pp. 105-113, 1995.

[37] U. Schlink, "On the probability model for asthma attacks," J. Theoretical Biol., Vol 215, pp. 405-414, 2002.

[38] W. R. Solomon, "Airborne pollen: a brief life," J Allergy Clin Immunol, Vol 109, pp. 895-900, 2002.

[39] W. R. Solomon, "How ill the wind? Issues in aeroallergen sampling," J Allergy Clin Immunol., Vol. 112, 3-8, 2003.

[40] B. R. Stern, M. E. Raizenne, R. T. Burnett, L. Jones, J. Kearney, and C. A. Franklin, "Air pollution and childhood respiratory health: exposure to sulfate and ozone in 10 Canadian rural communities," Environmental Research, Vol. 66, pp. 125-142, 1994.

[41] D. M. Stieb, R. T. Burnett, R. C. Beveridge, and J. R. Brook, "Association between ozone and asthma emergency department visits in Saint John, New Brunswick, Canada," Environ. Health Perspect., Vol. 104(12), pp. 1354-1360, 1996.

[42] J. Sunyer, R. Atkinson, F. Ballester, A. Le Tertre, J. G. Ayres, F. Forastiere, B. Forsberg, J. M. Vonk, L. Bisanti, R. H. Anderson, J. Schwartz, and K. Katsouyanni, "Respiratory effects of sulphur dioxide: a hierarchical multicity analysis in the APHEA 2 study," Occup Environ Med, Vol. 60, 2003.

[43] P. E. Taylor, R. C. Flagan, R. Valenta, and M. M. Glovsky, "Release of allergens as respirable aerosols: a link between grass pollen and asthma" J. Allergy Clin Immunol, Vol 109, pp. 51-56, 2002.

[44] J. M. Tenias, F. Ballester, M. L. Rivera, "Association between hospital emergency visits for asthma and air pollution in Valencia, Spain," Occup. Environ. Med., Vol. 55, pp. 541-547, 1998.

[45] S. D. Thomas, and S. Whitman, "Asthma hospitalizations and mortality in Chicago: an epidemiologic overview," Chest, Vol. 116, pp. 135S-141S, 1999.

[46] T. Beer, "Air quality as a meteorological hazard," Natural Hazards, Vol. 23, pp. 157-169, 2001.

[47] A. J. Venn, S. A. Lewis, M. Cooper, R. Hubbard, and J. Britton, "Living near a main road and the risk of wheezing illness in children," American J Respir Crit Care Med, Vol 164, pp. 2177-2180, 2001.

[48] D. J. Ward, and J. G. Ayres, "Particulate air pollution and panel studies in children: a systematic review," Occup. Environ Med, Vol 61, 2004.

[49] T. W. Wong, Y. T. Wun, T. S. Yu, W. Tam, C. M. Wong, and A. H. S. Wong, "Air pollution and general practice consultations for respiratory illnesses," J. Epidemiol. Community Health, Vol. 56, pp. 949-950, 2002.

[50] T. W. Wong, T. S. Lau, T. S. Yu, A. Neller, S. L. Wong, W. Tam, and S. W. Pang, "Air pollution and hospital admissions for respiratory and cardiovascular diseases in Hong Kong," Occup. Environ. Med., Vol. 56, pp. 679-683, 1999.

[51] A. Zeghnoun, P. Beaudeau, F. Carrat, V. Delmas, O. Boudhabhay, F. Gayon, D. Guincetre, and P. Czernichow, "Air pollution and respiratory drug sales in the city of Le Havre, France, 1993-1996," Environmental Research, Section A, Vol. 81, pp.224-230, 1999.

[52] R. Zhang, W. Lei, X. Tie, and P. Hess, "Industrial emissions cause extreme urban ozone diurnal variability," Proceed. National Acad. Sci., Vol. 101(17), pp. 6346-6350, 2004.

[53] L. H. Ziska, D. E. Gebhard, D. A. Frenz, S. Faulkner, B. D. Singer and J. G. Straka, "Cities as harbingers of climate change: common ragweed, urbanization, and public health," J. Allergy Clin Immunol, Vol 111, pp. 290-295, 2003.

[54] M. Zureik, C. Neukirch, B. Leynaert, R. Liard, J. Bousquet, and F. Neukirch, "Sensitisation to airborne moulds and severity of asthma: cross sectional study from European Community respiratory health survey," British Medical Journal, Vol. 325, pp. 411, 2002.

[55] P. J. Gergen, H. Mitchell, and H. Lynn, "Understanding the seasonal pattern of childhood asthma: results from the National Cooperative Inner-City Asthma Study (NCICAS)," J. Pediatics, Vol. 141, pp. 631-636, 2002.

[56] J. P. M. Braat, P. G. Mulder, H. J. Duivenvoorden, R. G. Van Wijk, E. Rijntjes, and W. J. Fokkens, "Pollutional and meteorological factors are closely related to complaints of non-allergenic, non-infectious perennial rhinitis patients: a time series model," Clin. Exp. Allergy, Vol. 32, pp. 690-697, 2002.

[57] J. Carrano, "Chemical and Biological Sensor Standards Study," DARPA, http://www1.va.gov/ vasafety/docs/C_B_Senor_Stand_Study_Report.pdf, 2004.

[58] http://www4.ncdc.noaa.gov/cgi-win/wwcgi.dll?wwDI~StnSrch~StnID~20027254#ONLINE

CHAPTER 7

AN EVALUATION OF OVER-THE-COUNTER MEDICATION SALES FOR SYNDROMIC SURVEILLANCE

Murray Campbell, Chung-Sheng Li, Charu Aggarwal, Milind Naphade
Kun-Lung Wu, Tong Zhang

IBM T.J. Watson Res. Ctr, 19 Skyline Drive, Hawthorne, NY 10532, USA

Abstract

Early and reliable detection of disease outbreaks is an important problem for public health. Syndromic surveillance systems use pre-diagnostic data sources to attempt to improve the timeliness of outbreak detection. This paper describes a number of approaches to evaluating the utility of data sources in a syndromic surveillance context. We show that there is some evidence that sales of over-the-counter medications have value for syndromic surveillance.

1. Introduction

Syndromic surveillance refers to the use of pre-diagnostic health-related data for early detection of disease outbreaks. With recent concern over the threat of bioterrorism, as well as the appearance of new disease threats (e.g., SARS), syndromic surveillance is being looked to as a means to improve the timeliness of public health surveillance.

The development of a useful syndromic surveillance system depends in part on the identification of data sources that have value in predicting disease outbreaks. This paper will focus on methods for assessing the value of data sources for predicting disease outbreaks. We will examine a number of different approaches that use retrospective analysis to evaluate data sources.

A frequently cited example of a data source that is presumed to be useful for syndromic surveillance is the sale of over-the-counter (OTC) medications [13,17,18]. We will apply our evaluation approaches to a large, multi-year, multi-city data set and show that there is some evidence that OTC medication sales may be useful for syndromic surveillance.

2. Background and Related Work

Syndromic surveillance (also referred to in the literature as early detection of disease outbreaks, pre-diagnosis surveillance, non-traditional surveillance, enhanced surveillance, non-traditional surveillance, and disease early warning systems) has received substantial interest recently, especially after Sept. 11, 2001 [3,5,10,14,15,16,17].

A number of studies have been devoted to investigating various data sources, such as the text and the ICD-9 diagnosis code of the chief complaints from emergency department [1,2,7,12], 911 call [4], and over-the-counter (OTC) drug sales [9].

There are at least three different classes of approaches to evaluating the utility of a data sources for syndromic surveillance. The first approach is based on the measuring the correlation between a target data source and a gold standard (diagnostic) data source [18]. A second approach is to use the target data source to better predict values in the gold standard data source. A third option is to identify "events" (i.e., disease outbreaks) in a gold standard data source, and assess the timeliness of alarms produced by a detection algorithm operating on the target data source. The tradeoff between timeliness and false alarms can be assessed using the AMOC approach [8].

3. Data

There are two data sets that will be used in this study. The first, which we will call OTC, is a weekly summary of unit sales of upper respiratory over-the-counter medication sales for ten cities (Baltimore/Washington, Charlotte, Chicago, Dallas, Milwaukee, New York, Norfolk, Orlando, Pittsburgh, and Seattle) for a three-year period (2000-2002). The first data point is for the week ending on 1/9/2000, and the last data point is for the week ending 12/29/2002. For each city, sales are reported in eight categories: four types (Cold, Allergy, Cough, and Sinus), and two target groups for each type (Adult and Pediatric).

The second data set, which we will call CL, consists of anonymized medical insurance claims records corresponding to physician office visits. The records are from the same ten cities as for OTC, and cover the same three-year period. Each record consists of a unique (anonymized) patient identifier, a date of service, up to four ICD-9 (diagnosis) codes, and a city name. There are a total of about 22.5 million records. The ICD-9 codes were chosen by the data provider, Surveillance Data, Inc., to be relevant to upper respiratory infections. The number of insurance claims was aggregated by city to weekly totals aligned with the OTC data.

For the purposes of this study, the OTC data set is the target data source, i.e., OTC will be assessed for value in syndromic surveillance. CL is the gold standard data source, as it contains diagnostic information about actual disease.

4. Approaches

4.1. *Lead-lag correlation analysis*

One approach to evaluating a data source for syndromic surveillance is to conduct a lead-lag correlation analysis on the data source with respect to a gold standard data source. This consists of computing the correlation between the two time series for a range of lead-lag times, and identifying the lead-lag time at which the correlation is maximized. It can be useful to remove trends before analyzing.

Although a correlation analysis can give a global view of the lead time of a target data source, syndromic surveillance is typically more interested in the lead time prior to increasing levels of disease. This suggests an alternative approach where a correlation analysis is performed on a number of shorter time segments that contain the initial stages of disease outbreaks.

In Section 5.1 we will apply this method to the data sets described in Section 3, and assess the value of OTC data for syndromic surveillance.

4.2. *Regression test of predictive ability*

This section describes another approach to evaluating the usefulness of a target data source by posing it as a prediction problem. More specifically, we are interested in predicting certain quantities associated with the gold standard data source, and want to see whether by including the target data, we are able to make better predictions.

This approach can be generally regarded as time-series forecasting. If we can forecast a quantity A more accurately using a quantity B under a certain metric, then we say that B contains useful information for predicting A.

We now give a general description of this approach. Assume that the quantity of interests is presented sequentially as a time-series

$$\{Y\} = \{\cdots, Y_0, Y_1, \cdots, Y_t, \cdots\} \tag{1}$$

We want to predict the future values of this time-series based on some side-information (which may include the historical values of Y we observed so-far), represented as another time-series of vectors:

$$\{X\} = \{\cdots, X_0, X_1, \cdots, X_t, \cdots\} \tag{2}$$

Each Y_t is a real-valued number, observed at time t, which we are interested in. Each X_t is a real-vector, which encodes all of the side information that we hope are useful for predicting the $\{Y\}$ series.

To this end, we assume that at each time t, based on the current side-information X_t, we would like to predict Y_{t+f}, which is the value of the Y series f steps in the future (where $f > 0$ is an integer). We assume that the predictor $p_f(X_t)$ has a linear form as

$$Y_{t+f} \approx p_f(X_t) = w_f^T X_t \tag{3}$$

where w_f is a weight vector (parameter of our model) that characterizes the predictor p_f. The parameter w_f can be estimated from the data (as we will describe later).

Given a predictor, represented as a weight vector $\mathbf{w}$, we can measure its quality using a certain figure of merit. In this study, we employ the commonly used least-squares error criterion, defined as

$$R_f(w,[T_1,T_2]) = \frac{1}{T_2 - T_1 + 1} \sum_{t=1}^{T-f} (w^T X_t - Y_{t+f})^2 . \qquad (4)$$

The number $R_f(w,[T_1,T_2])$ measures in the interval $[T_1,T_2]$ how well we can predict from X the sequence Y f steps in advance with the weight vector w.

The weight vector can be estimated from the historical data using least-squares regression:

$$\hat{w}_{f,T} = \arg \min_w \sum_{t=1}^{T-f} (w^T X_t - Y_{t+f})^2 . \qquad (5)$$

Now assume that we observe the sequences X and Y, up to some point T. To check how useful X is for predicting Y, we divide the time period into K consecutive blocks (for simplicity, assume that T is divisible by K): $I_j = [T_j, T_{j+1}]$ for $j = 0,\dots,K-1$, where $T_j = hT/K$. Now we can use a single number

$$r_f(X,Y) = \frac{1}{K} \sum_{j=1}^{K-1} R_f(\hat{w}_{f,T},[T_j,T_{j+1}]) \qquad (6)$$

to measure the usefulness of X for predicting Y (f steps in the future). That is, we train a predictor $\hat{w}_{f,T_j}$ using least squares regression [Eq. (5)] with data observed up to jT_0, and then test on data from jT_0 to T_{j+1}, for $j = 0,\dots K-1$, and then average the results. The smaller $r_f(X,Y)$ is, the more useful X is for predicting Y. Therefore using Eq. (6), we can compare the usefulness of different side informations X and X'.

In Section 5.2, we compute the corresponding $r_f(X,Y)$ numbers with and without including the OTC data in the side information X. Our results suggest the usefulness of the OTC data in public health surveillance.

4.3. *Detection-based approaches*

For the detection-based approaches we assume that disease outbreak events are labeled in the gold standard data set, and an outbreak detection algorithm operates on either the target data set or the gold standard data set. Using the AMOC approach, we are able to assess the lead time provided by the target data source over a range of practical false alarm rates.

4.4. *Supervised algorithm for outbreak detection in OTC data*

The supervised [6] outbreak detection algorithm utilized the previously supplied data in order to determine various aspects of the algorithm. The supervised algorithm required a number of components in order to perform the detection:

 (i) Determination of features to be used, and the proper way to combine channels.
 (ii) Creation of streams of anomalies.
 (iii) Conversion of the anomaly streams into the alarm level using the information from (1).

This supervision was done in two forms:

Feature Selection: Since multiple channels of information were available, which channels provided the greatest level of connection between the channels and actual outbreaks?

Combination of Multiple Channels: How do we combine the signals from multiple channels in order to create one integrated alarm level which was most effective for detecting the outbreak?

In order to perform feature selection, we used the same OTC data set (provided by SDI) as described in the other sections. The first step was to determine which of the channels were most discriminatory for the purpose of distinguishing the biological outbreak from the background noise.

Let us assume that for each site i, the value indicating the channel specific information (absentee behavior, phone calls, pharmacy buying behavior) at time t is denoted by $y(i,t)$. The first step was to convert the data into statistical deviation levels which could be compared across different features. Thus, each stream of data was converted into a statistical stream of numbers indicating the deviation level with respect to the prior window of behavior of width W. The statistical deviation value for a given stream i at time t was denoted by $z(i,t)$. The value of $z(i,t)$ was found by first fitting the prior window of width W with the polynomial function $f(t)$. The deviation value at time t_0 was then defined as follows:

$$s(i) = \sqrt{\sum_{t=t_0-W}^{t_0} (f(t) - y(i,t))^2 / (W - 1)} \qquad (7)$$

The value of W used was based on the last 16 reports. This statistical deviation is also referred to as the z-number. This value provides an idea of how far the stream of data deviates from the normal behavior and gives an intuitive understanding of the level of anomaly at a given tick. Then, the statistical deviation $z(i,t_0)$ at time t_0 is denoted by:

$$z(i,t_0) = (f(t_0) - y(i,t_0)) / s(i) \qquad (8)$$

These alarm values could be used in order to determine the value of each channel in the training data. A particular channel was found to be useful when this value was found to be larger than a pre-defined threshold of 1.5. For example, by using this technique we were able to eliminate the allergy channel for the purpose of detection of the flu infections. For example, this behavior was illustrated by the allergy channel in the OTC training data. We have also illustrated the AMOC curve for the allergy channel in the same figure. We note that the AMOC curve for the allergy channel was particularly poor, because it seemed to be uncorrelated to the seasonal outbreaks in the data.

Once these features were selected, they could be used on the test data for computing the statistical deviation values using the same methodology as discussed above. Thus, a separate signal was obtained from stream. The next step was to combine the deviation values from the different sites and channels to create one composite signal. A supervised training process was utilized to determine the optimal functional form for the test data. This was achieved by finding the composition which maximized the area under the AMOC curve.

Once each channel had been converted into a single composite signal, they need to be combined to create a combination signature. For example, let $q_1(t)$, $q_2(t)$ and $q_3(t)$ be the signatures obtained from three different channels. The combination signature was defined as the expression:

$$C(t) = c_1 \cdot q_1(t) + c_2 \cdot q_2(t) + c_3 \cdot q_3(t) \tag{9}$$

Here c_1, c_2, and c_3 were coefficients which were also determined by minimizing the latency of detection on the training data. As a normalization condition, it is assumed that the coefficients satisfy the following condition for the constant C':

$$c_1^2 + c_2^2 + c_3^2 = C' \tag{10}$$

It is necessary to use the above condition for scaling purposes. In order to determine the optimal alarm we found values of c_1, c_2, and c_3, which optimized the area under the AMOC curve. This provides the combination signature.

4.5. *Modified Holt-Winters forecaster*

One of the unsupervised detectors used was a modified Holt-Winter forecaster [11]. The forecaster generated a z-value for each tick of a data channel, representing the deviation of observed data from the predicted one. A z-value is computed as follows:

$$z = (\Delta - \mu)/\sigma \tag{11}$$

where Δ is the difference between observed and predicted data, and μ and σ are the mean and standard deviation, respectively, of these Δ differences in the past.

A Holt-Winters forecaster assumes that a time series $X_1, \cdots, X_N$ can be modeled in terms of three key components: the average X_N, the trend T_N and the daily seasonality factors $F_{N-D+1}, \cdots, F_N$, where D is the number of days in the week for which there are observed data. The average is the exponentially smoothed level value of all the time series values. The trend is the exponentially smoothed slope of all the N time series values. The daily seasonality factors are exponentially smoothed values reflecting the deviation from linearity attributable to the different days of the week. The seasonality factors can have either a multiplicative or additive effect. In our implementation, we chose the additive variant. A Holt-Winters forecaster attempts to accurately capture these three key components of a time series. It can deal with special events, such as holidays or special days where data are missing.

4.6. *Forecasting based on multi-channel regression*

A simple prediction strategy that can combine single and multi-channel prediction is to set up the problem as a linear regression. As usual, the deviation of the actual value from the predicted value is a measure of abnormality. We set up a system of linear equations as shown below.

Let the observation stream of a single channel from among the multiple OTC sales channels be $[y_1, ..., y_M]$. Consider using the past J observations to derive the regression parameters while using the past K samples for actually predicting the $K+1^{th}$ observation. The number of variables to be estimated from the past J samples is K.

$$
\begin{bmatrix} y_{M-1} \\ y_{M-2-1} \\ \cdots \\ y_{M-J-1} \end{bmatrix} = \begin{bmatrix} y_{M-2} & \cdots & y_{M-K-1} \\ y_{M-3} & \cdots & y_{M-K-2} \\ \cdots & \cdots & \cdots \\ y_{M-J-2} & \cdots & y_{M-K-J-1} \end{bmatrix} \begin{bmatrix} w_1 \\ w_2 \\ \cdots \\ w_K \end{bmatrix} \tag{12}
$$

or using matrix notation:

$$
Y = A_y W \tag{13}
$$

With this overdetermined system of equations $(J > K)$ we then calculate the least squares fit to this as shown in Eq. (14):

$$
W = (A_y^t A_y)^{-1} A_y^t Y \tag{14}
$$

Assuming linear independence among columns of matrix A, $A^t A$ is non-singular and the generalized inverse $(A^t A)^{-1}$ exists. We calculate the weight vector W after every update. Thus for each observation y_M we calculate the prediction aW, a being a row vector $\begin{bmatrix} y_{M-1} & y_{M-2-1} & \cdots & y_{M-J-1} \end{bmatrix}$. If the residual between the actual value and the predicted value is positive we use this difference as a measure of abnormality and probability of an outbreak. Equation 12 can be extended to make the prediction based on multiple channels. For example the matrix A can be created by combining multiple channels. Equation 15 shows past samples from two channels $[y_1,...,y_M]$ and $[x_1,...,x_M]$ being used to predict the current observation of channel Y.

$$
Y = [A_y A_x] \begin{bmatrix} W_y \\ W_x \end{bmatrix} \tag{15}
$$

Using the above formulation we can predict the current value of sales of any of the OTC channel based on values of sales in the same channel as well as based on values of sales in additional channels.

5. Experiments

5.1. *Lead-lag correlation analysis of OTC data*

The lead-lag correlation analysis approach requires us, for each city, to compute the correlations corresponding to various possible lead-lag times. In Fig. 1, we examine offsets ranging from five weeks prior to five weeks after. The ten solid lines are the correlation values for each of the ten cities. The dashed line is the mean of those values. The peak correlation is between one and two weeks leading, i.e., OTC leading CL by one to two weeks. If a quadratic is fitted to the dashed line, the maximum is at 1.7 weeks.

This provides evidence, albeit somewhat weak, that OTC leads CL and may have value for syndromic surveillance. Clearly there is a wide discrepancy on the correlation between OTC and CL across the different cities, and this needs further investigation.

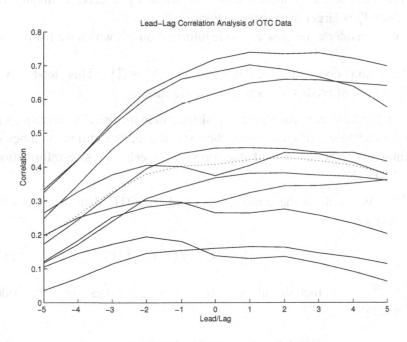

Figure 1: Lead-lag correlation analysis experiment.

5.2. *Regression test of the predicative value of OTC*

We study the usefulness of OTC for predicting insurance claims using the approach described in Section 4.2. Since the OTC data are weekly based, we shall form the time series on a weekly basis. In particular, we convert the insurance data into weekly data aligned with the OTC data.

In this experiment, we consider different cities separately. That is, we do not consider possible inter-city correlations. For each city, we let OTC_t be the total number of OTC sales in week t, and CL_t be the number of insurance claims in week t. Since in public health surveillance, we are mostly interested in sudden outbreaks of diseases, we are interested in the log-ratio of the number of insurance claims in consecutive weeks. That is, at week t, the Y variable is given by

$$Y_t = \log_2(CL_t / CL_{t-1}).\qquad(16)$$

One may also use other quantities, such as whether the insurance claims next week is higher than this week by a certain amount (or whether Y_t is larger than some threshold).

We consider a few possible side information X, which we list below.

X^1: Using constant side information: $X_t^1 \equiv [1]$. This leads to a predicator that predict Y_t using its historic mean.

X^{CL}: In addition to the above, we also include historical observations of the insurance claim data itself (the log ratio of the current number of claims over the claims of the previous week) as side-information: $X_t^{CL} = [Y_t, 1]$.

X_t^{OTC}: We include the constant one and the OTC data into the side-information:

$$X_t^{OTC} = [\log(OTC_t / OTC_{t-1}), 1].\qquad(17)$$

X_t^{CL-OTC}: We include all of the above quantities into the side-information:

$$X_t^{CL-OTC} = [\log_2(OTC_t / OTC_{t-1}), Y_t, 1].\qquad(18)$$

Since this framework is quite flexible, various other configurations can also be studied. For our purpose, we are able to make interesting observations from this particular configuration. Variations will lead to similar results.

Applying the notation in Section 4.2, for each city, we divide the time series into $K = 20$ blocks, and compute the $r_f(X,Y)$ number in Eq. (6) for $f = 1,2$ and each side information listed above. We then average the results over the ten cities, and report the averaged numbers in Table 1. From the table, we can see that the OTC data has a small predicative power for the insurance claims data CL. One may also do an experiment in the reverse order (that is, use historical CL data to predict the future OTC sales). In this case, for $f = 1$, the predictive performance for OTC sales, measured by the r_f value, degrades from 0.0217 (without CL in the side-information) to 0.0221 (with historical CL data in the side-information). Therefore these experiments provide some evidence suggesting that OTC changes precede CL changes.

Table 1. Averaged $r_f(X,Y)$ numbers over ten cities.

	X^1	X^{CL}	X^{OTC}	X^{CL-OTC}
$f = 1$	0.0287	0.0265	0.0285	0.0261
$f = 2$	0.0287	0.0291	0.0280	0.0287

Although effects shown in Table 1 are relatively small, we believe they are still indicative statistically. Since we average our results over ten cities, we may also check the variation over different cities. In particular, in seven out of ten cities, $r_1(X^{OTC},Y)$ is smaller than $r_1(X^1,Y)$; also in seven out of ten cities, $r_2(X^{OTC},Y)$ is smaller than $r_2(X^1,Y)$. This comparison is consistent with results in Table 1, and justifies from a slightly different point of view that statistically, the OTC data is (weakly) useful for predicting future insurance claims.

5.3. *Results from detection-based approaches*

5.3.1. *Supervised method*

Once the features have been selected, and the proper way for construction of the combination signature was determined, the actual alarm level construction on the data is straightforward. The deviation values for the data are computed in an exactly identical way to the training data, and the combination is created to output the corresponding alarm levels at each tick. In Fig. 2, we have illustrated the behavior of the detection algorithms. One interesting observation is that the OTC data is always more effective than the claims data. In fact, in most cases, the OTC data acted as a "leading indicator" over the claims data. It is also interesting to note that the adult and pediatric data illustrate differential behavior in terms of the speed and quality of the detection. An example of this is illustrated in Fig. 3.

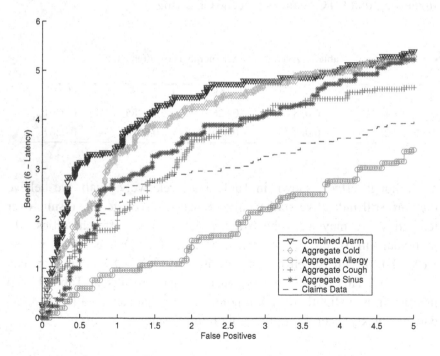

Figure 2: The AMOC curves generated by the Supervised method illustrate that various OTC categories are more timely than claims.

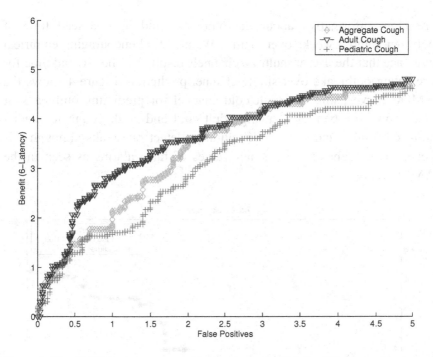

Figure 3: The AMOC curves generated by the Supervised Method illustrates that there is differentiation between Adult and Pediatric cough medication sales.

5.3.2. *Modified Holt-Winters forecaster*

Even though the OTC data are weekly data, the detector treats them as daily data and assumed that there were 3 days in a week. It uses the past 6 OTC data points to predict the next OTC sale.

While there was some variability across different categories of OTC medication sales, over a wide range of false alarm rates the Holt-Winters forecaster shows a two week lead time for OTC over Claims. Sinus medication sales are observed to have the best lead times overall.

5.3.3. *Forecasting based on multi-channel regression*

Using the OTC data we experimented with different values of J and K (see Section 4.6 for single and multi-channel prediction based outbreak detection. Based on our experiments we found that sales of adult drugs are more informative about the outbreaks and have a lead time of between 2 and 3 weeks over claims. We also find encouraging empirical evidence that the use of multiple channels results in a better lead time for predicting outbreaks over single channel prediction. Figure 4 shows the AMOC curve using the adult cold channel for predicting outbreaks. It also shows the benefit of using adult cold and adult cough to predict adult cold sales and use the deviation to detect outbreaks although this benefit is evident only for small values of false alarms as seen in the AMOC curve.

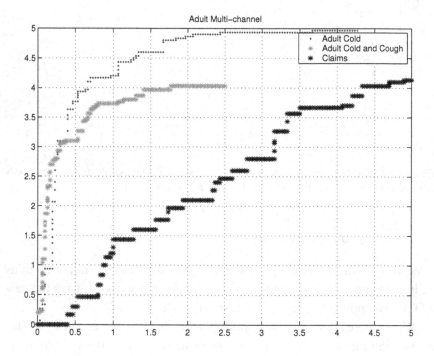

Figure 4: The Adult Cold sales are found to be the best indicator for the outbreaks with J=15, K=2 and J=20, K=1 respectively for single channel and multi-channel prediction. The Adult Cold and Cough sales are used in the two channel prediction.

6. Conclusions and Future Work

We have shown a number of different approaches to assessing the value of a data source for syndromic surveillance, and evaluated over-the-counter medication sales using these approaches. There appears to be evidence from each of these approaches that OTC medication sales are a leading indicator for disease outbreaks.

There are a number of limitations in this study. The data sets were aggregated weekly, which reduces the precision regarding estimates of the timeliness of OTC. This type of study should be repeated with daily data. The detection-based experiments identified only those outbreaks that occurred at the beginning of the seasonal rise in respiratory disease. A more careful study could examine finer grain disease outbreaks, preferably those that have been studied and verified by public health. This study was retrospective, looking only at historical data. A prospective study, using the target data source to predict disease outbreak in real time, would provide greater confidence in the conclusions in this chapter.

Acknowledgments

We would like to thank Andrew Kress of Surveillance Data, Inc. for supplying the data used in this study. This work was supported by the Air Force Research Laboratory (AFRL)/Defense Advanced Research Projects Agency (DARPA) under AFRL Contract No. F30602-01-C-0184. Any opinions, findings, conclusions, or recommendations expressed in this material are those of the authors and do not necessarily reflect the views of the AFRL or DARPA.

References

[1] E. M. Begier, D. Sockwell, L. M. Branch, J. O. Davies-Cole, L. H. Jones, L. Edwards, J. A. Casani, and D. Blythe. The national capitol region's emergency department syndromic surveillance system: do chief complaint and discharge diagnosis yield different results? *Emerging Infectious Disease*, 9(3):393--396, Mar. 2003.

[2] J. Beitel, K. L. Olson, B. Y. Reis, and K. D. Mandl. Use of emergency department chief complaint and diagnostic codes for identifying repiratory illness in a pediatric population. *Pediatric Emergency Care*, 20(6):355--360, Jun. 2004.

[3] J. W. Buehler, R. L. Berkelman, D. M. Hartley, and C. J. Peters. Syndromic surveillance and bioterrorism-related epidemics. *Emerging Infectious Diseases*, 9(10):1197--1204, Oct. 2003.

[4] M. R. Dockrey, L. J. Trigg, and W. B. Lober. An information systems for 911 dispatch monitoring system and analysis. In *Proceeding of the AMIA Symposium*, page 1008, 2002.

[5] J. S. Duchin. Epidemiological response to syndromic surveillance signals. *Journal of Urban Health*, 80(2):i115--i116, 2003.

[6] R. O. Duda, P. E. Hart, and D. G. Stork. *Pattern Classification*. John Wiley & Sons, Inc., 2001.

[7] J. U. Espino and M. M. Wagner. Accuracy of icd-9-coded chief complaints and diagnoses for the detection of acute respiratory illness. In *Proceedings AMIA Symposium*, pages 164--168, 2001.

[8] T. Fawcett and F. Provost. Activity monitoring: Noticing interesting changes in behavior. In Chaudhuri and Madigan, editors, *Proceedings on the Fifth ACM SIGKDD International Conference on Knowledge Discovery and Data Mining*, pages 53--62, San Diego, CA, 1999.

[9] Goldenberg, G. Shmueli, R. A. Caruana, and S. E. Fienberg. Early statistical detection of anthrax outbreaks by tracking over-the-counter medication sales. *Proceedings of the National Academy of Sciences of the United States of America*, 99(8):5237--5240, Apr. 2002.

[10] T. Goodwin and E. Noji. Syndromic surveillance. *European Journal of Emergency Medicine*, 11(1):1--2, Feb. 2004.

[11] Granger and P. Newbold. *Forecasting Economic Time Series*. Academic Press, 1977.

[12] J. Greenko, F. Mostashari, A. Fine, and M. Layton. Clinical evaluation of the emergency medical services (ems) ambulance dispatch-based syndromic surveillance system, New York City. *Journal of Urban Health*, 80(2):i50--i56, 2003.

[13] W. R. Hogan, F. C. Tsui, O. Ivanov, P. H. Gesteland, S. Grannis, J. M. Overhage, J. M. Robinson, and M. M. Wagner. Detection of Pediatric Respiratory and Diarrheal Outbreaks from Sales of Over-the-counter Electrolyte Products. *Journal of the American Medical Informatics Association*, 10(6):555-562, 2003.

[14] F. Mostashari and J. Hartman. Syndromic surveillance: A local perspective. *Journal of Urban Health*, 80(2):i1--i7, 2003.

[15] J. A. Pavlin. Investigation of disease outbreaks detected by syndromic surveillance systems. *Journal of Urban Health*, 80(2):i107--i114, 2003.

[16] M. Sosin. Syndromic surveillance: The case for skillful investment. *Biosecurity and Bioterrorism: Biodefense Strategy, Practice, and Science*, 1(4):247--253, 2003.

[17] M. M. Wagner, F. C. Tsui, J. U. Espino, V. M. Dato, D. F. Sittig, R. A. Caruana, L. F. McGinnis, D. W. Deerfield, M. J. Druzdzel, and D. B. Fridsma. The emerging science of very early detection of disease outbreaks. *Journal of Public Health Management Practice*, 7(6):51--59, Nov. 2001.

[18] R. Welliver, J. Cherry, K. Boyer, J. Deseda-Tous, P. Krause, J. Dudley, R. Murray, W. Wingert, J. Champion, and G. Freeman. Sales of nonprescription cold remedies: a unique method of influenza surveillance. *Pediatric Research*, 13:1015--1017, 1979.

CHAPTER 8

COLLABORATIVE HEALTH SENTINEL

J. H. Kaufman, G. Decad, I. Eiron, D. A. Ford, D. Spellmeyer,[1] C-S. Li[2], and
A Shabo (Shvo)[3]

*IBM Research Division [1]Almaden Research Center, [2]Watson Research Center,
[3]Haifa Research Lab*

1. Introduction

The rise of global economies in the 21st century, and the increased
reliance by developed countries on global transportation and trade,
increases the risk of national and worldwide disaster due to infectious
disease. Such epidemics may be the result of global climate change,
vector-borne diseases, food-borne illness, new naturally occurring
pathogens, or bio-terrorist attacks. The threat is most severe for rapidly
communicable diseases. The coincidence of rapidly spreading infectious
disease along with the rapid transportation, propagation, and
dissemination of the pathogens and vectors for infection poses the risk of
new and dangerous pandemics [1,2,3].

The anthrax attacks in the United States [4], which occurred shortly
after the events of September 11th, demonstrated the subtle nature of
biological attacks and their effectiveness in spreading terror [5]. The
effects of that particular attack were mitigated by the fact that anthrax is
not easily weaponized and the infections were limited by physical
distribution of the anthrax spores (i.e. no person-to-person transmission)
[6]. Bio-terrorist attack based on other agents could be significantly
more dangerous and difficult to contain [4–7]. One of the significant
lessons to be learned from that attack is the difficultly of early
recognition of biological threats [5].

The requirements for reporting diseases in the United States are mandated by state laws or regulations, and the list of reportable diseases in each state differs. CDC provides a uniform criteria for reporting cases [8]. The Center for Disease Control (CDC) maintains a list of Nationally Notifiable Diseases considered being significant threats to pubic health. Individual states also define lists of reportable diseases. As a result of this protocol, the first line of defense for developed nations against infectious disease is our world-class medical care infrastructure. In the United States, physicians depend on several important programs, many administered by the CDC [9], to receive early warning and prepare to respond to new epidemics. To date, these programs have proven effective in protecting society from naturally occurring pathogens and food borne illness [10–16]. However, many current programs are optimized to help physicians diagnose or recognize new illnesses [17]. The data these programs depend upon come, in large part, from local and regional medical institutions, laboratories, and insurance companies [10–22]. Therefore, our current bio defense infrastructure is optimized to detect illness after people have shown clinical symptoms and have gone for medical help. Since expression of such symptoms often requires an incubation time of several days, infected persons with sub-clinical symptoms can be propagating new infections for days before early warning is sent by medical institutions.

The vulnerability of our society to fast spreading naturally occurring agents has also been demonstrated in recent years, e.g., in the SARS outbreak [23], cases of West Nile Virus [24], and Dengue Fever [25]. Early detection of infectious disease is both essential to public health efforts and a vital first step in the fight against bio-terrorism.

The need for an early warning system has been recognized and a number of local or regional experimental programs are now in place [26–29]. These programs often depend on hospital emergency rooms for data and critics argue this puts an undue burden on our healthcare infrastructure. Nonetheless, the need for a national real-time health monitoring system is well recognized [26–29].

The time is right for our society to put in place a national early warning system to guard against the onset of dangerous infectious disease – whether triggered by hostile forces or naturally occurring

[1–3, 21]. The goal of such a system should not be to diagnose individuals and their specific ailments, but to rapidly identify the existence, nature, severity, and location of potential threats [26–29]. This new system must be able to "sound the alarm" early in the nucleation stage before the pathogens result in large numbers of clinical patients. In order to accomplish this, in direct analogy to weather forecasting, the system must gather data nationally and with fairly high resolution. A fine scale local grid in New York City will not detect a case of SARS in Denver, CO. The raw data sources should be real time and, ideally, should originate while symptoms are still pre-clinical. Rather than rely on indirect measurements or time-delayed indications of symptoms, the primary data should be provided directly by a core sample group of individuals. This is not to understate the role of hospitals and physicians. Instead, we propose traditional diagnostic data can be augmented by information obtained directly from this national sample population. This will provide earlier warning and reduce the burden on healthcare provider. Previous (small scale) experiments have demonstrated the usefulness of a collaborative health sentinel as an early warning system. EpiSPIRE provided a health monitoring system for 2002 flu season for a population of scientists at IBM Watson Research Center [30].

In this chapter we propose the formation of the Collaborative Health Sentinel (CHS). CHS is a system and protocol designed to provide early detection of infectious agents. CHS relies on our existing infrastructure for diagnosis and treatment planning. CHS works by establishing a longitudinal population health baseline through which an understanding of the dynamic national health "weather map" may be achieved. Today's information infrastructure can make such a system cost effective. Widespread distribution of computerized communications technology to the general population makes creation of this network technically feasible and practical. We propose a data collection system where a core group of individuals (representative of the underlying population) provide regular health data without involving medical caregivers. This primary data source will be used to generate a national real time health map. There are potentially hundreds or thousands of attributes that could be used to construct the so-called health map. CHS relies on well documented descriptions of CDC reportable diseases to define the raw data types collected directly from the core group of reporting individuals. It also provides tools for researchers to study and identify the relevant parameters, correlations, and models, required to transform the symptomatic health map into a real-time "risk" map for reportable diseases.

CHS will also provide a standards based interface to health care providers so they will have early access not only to the health map (the big picture of emerging epidemics) but also to specific data relevant to treating individuals in need of care. The interface is bi-directional so health care providers may input new clinical data (and even diagnoses) to refine the information available in CHS for the benefit of others. In analogy to weather forecast, it will be necessary to develop new models to project foreword the progress of disease and to make forecasts. Accordingly, the CHS health map will provide an interface and tools for experts to develop their own forecast models based on the CHS data and using several approaches including, dynamical systems analysis, epidemiological models, and data mining and pattern matching.

2. Infectious Disease and Existing Health Surveillance Programs

Infectious disease is usually caused by the transmission of pathogens directly between humans (such as flu and smallpox) or through vectors as in the case of vector-born disease (such as mosquitoes, rodents, etc.). Vector-born epidemic disease, such as Hantavirus Pulmonary Disease, Dengue fever, malaria, St. Louis encephalitis, Plague, Yellow fever, and West Niles encephalitis, can be transmitted through insects or rodents [9]. Several factors influence the transmission of this type of disease including urbanization, deforestation, and agriculture practices. All of these vector-born diseases are modulated by environmental factors such as temperature and moisture patterns and geographical factors. Some of the vector-born disease may also affect other types of animals (such as West Niles encephalitis on crow).

In contrast, infectious diseases can also be a result of biological warfare, which is defined as the release of bio-agents and pathogens (such as anthrax) through either aerosol or water-born mechanisms to inflict either targeted or untargeted casualties on human beings. These bio-agents or pathogens released as a result of biological warfare, bio-terrorism, or bio-crime are indistinguishable from those pathogens having natural cause. The transmission mechanisms of these pathogens include:

- Inhalation, direct contact, or through eating,
- Single release of the pathogen to a densely populated location through aerosol mechanism, such as during the Super Bowl game, or to a reservoir which provides water to a densely populated city,
- Some of the pathogens can be transmitted directly from human to human, while the others can only be transmitted through carriers (or vectors) and are thus called as vector-born disease

The United States has several strategic national programs designed to protect our population. Many of these "health surveillance" programs are administered by the Center for Disease Control and Prevention (CDC) [9]. Most of the data monitored by these programs is collected by medical professionals at hospitals and is, therefore, available only after people begin to show clinical symptoms and seek medical care. The primary focus of our national health surveillance programs has been diagnosis (especially of diseases not yet common in the United States) and communication of treatment strategies. We summarize some of the relevant programs below.

The CDC Public Health Information Network [17] (PHIN) leverages five components for detection and monitoring, data analysis, knowledge management, alerting and response. PHIN also helps define and implement information technology standards for public health (building on the technologies developed for the Health Alert Network [10] [HAN], National Electronic Disease Surveillance System [11] (NEDSS), and the Epidemic Information Exchange [12] (Epi-X). NEDSS and HAN use data standards derived from, and compatible with the Health Level 7 Reference Information Model [13] (RIM). The CDC PulseNet [14] program is a national food borne disease surveillance network designed to identify outbreaks of food borne illness. The CDC also issues "The Morbidity and Mortality Weekly Report" (MMWR) which provides statistics on a number of threats including infectious and chronic diseases reported by local health departments. The CDC Epidemic Intelligence Service [15] (EIS) Program helps local health officials identify new and emerging threats. EIS is a post-graduate program of service and on-the-job training for health professionals interested in the practice of epidemiology.   EIS also establishes risk factors for infection, and recommend control measures. In 1999, EIS officials identified cases of West Nile encephalitis. The CDC also publishes online The Emerging Infectious Disease Journal [16] intended to increase awareness of important public health information so prevention measures can be more rapidly implemented.

The **Laboratory Response Network** [18] (LRN) "…and its partners maintain an integrated national and international network of laboratories that is fully equipped to respond quickly to acts of chemical or biological terrorism, emerging infectious diseases and public health threats and emergencies."

The Food Emergency Response Network [19] (FERN) is "…a network of state and federal laboratories that are committed to analyzing food samples in the event of a biological, chemical, or radiological terrorist attack in this country."

The **Electronic Laboratory Exchange Network** [20] (eLEXNET) is "… a seamless, integrated, web-based information network that allows health officials at multiple government agencies engaged in food safety activities to compare, share and coordinate laboratory analysis findings. eLEXNET provides the necessary infrastructure for an early warning system that identifies potentially hazardous foods and enables health officials to assess risks and analyze trends."

The **National Animal Health Laboratory Network** [22] (NAHLN) is "…a network of federal and state resources intended to enable a rapid and sufficient response to animal health emergencies, including foot-and-mouth disease and other foreign animal diseases…. Its primary objective is to establish a functional national network of existing diagnostic laboratories to rapidly and accurately detect and report animal diseases of national interest, particularly those pathogens that have the potential to be intentionally introduced through bio-terrorism."

There is a growing interest in identifying environmental factors that contribute to diseases and other public health risks including Hantavirus, dengue fever, cholera, lyme disease and air pollution. Most of the existing epidemiological techniques rely on either passive or active surveillance of clinical data from health care facilities to identify outbreaks. In *passive surveillance*, health care providers, hospitals, and sometimes labs send reports to the health department as prescribed in a set of rules or regulations. In contrast, health department staff call or visit healthcare providers on a regular basis (e.g., weekly) to solicit case reports in *active surveillance*. An *outbreak* is then determined from the difference between the actual and expected number of new cases for a specific disease.

There is also recent movement towards analyzing environmental factors such as weather, landscape, and topography that influence the spread of epidemic diseases. Some of this information, such as weather and landscape data, is captured in remotely sensed data. The data is gathered on a continuous and wide-scale basis, and thus provides better coverage for predicting impending outbreaks than current passive surveillance techniques.

Several local and regional new efforts are being initiated to provide bio-surveillance [26–28]. "Essence" (developed by JHU/APL) has been in place to monitor the capital region of Baltimore/DC since 1999 [26]. The Real-time Outbreak and Disease Surveillance (RODS) [28] open source project is a system developed by U. of Pittsburgh to enable any state and local government to adopt a computer-based public health surveillance system for early detection of disease outbreaks. In the RODS design, Hospitals send data to a RODS database. RODS automatically classifies the main "complaints" into one of seven syndrome categories using Bayesian classifiers. The data is available for analysis using data warehousing techniques. The CDC has established an experimental http://www.cdc.gov/ "BioSense" surveillance system [9] to look for early signs of a bioterror attack. "BioSense" monitors a variety of data sources include over-the-counter drug sales and illnesses reported in hospital emergency rooms. These importance efforts reflect the recognized need for national health surveillance system.

In order to provide early warning of health threats the government has also initiated several efforts to exploit networks of sensors. Sensor networks can be designed to detect toxic chemicals, explosive and/or radiological agents, as well as biological agents that could threaten national security. Pervasive ultra-sensitive sensors positioned in key locations and/or on personnel involved in public service, travel and safety could provide an early warning of chemical and radiological agents and promote public safety by avoidance or evacuation of affected areas. Current programs are now in place to investigate the feasibility of locating sensors in key transportation hubs, high-rise office and apartment buildings, large public gathering areas, etc. The sensors could provide real-time monitoring and feedback to action-related public safety agencies. Oak Ridge National Lab has proposed a "SensorNet" to provide such detection including explosive threats (www.sensornet.gov). These systems of integrated sensors with state of the art nanotechnology may be able to detect ultra low levels of chemical and radioactive agents. NIST has been developing sensing devices using tin oxide as a base material with catalytic metal additives for chemical selectivity. Sandia Labs is developing microchemical sensors for volatile contaminants, "chemiresistors," (www.sandia.gov).

3. Elements of the Collaborative Health Sentinel (CHS) System

In the sections below, we discuss six important elements of the CHS system. The first is **Sampling**. The section defines the data the system must collect, who should generate the data, and how should it be collected. Secondly, we discuss how the data can be used to create a real time **National Health Map**. Thirdly, we describe how the CHS health map may be used to **Detect** emerging patterns of infection. Fourth, we discuss how CHS interacts with other programs to **React** to these events. Finally, we consider the system **Cost** and cost effective design.

3.1. *Sampling*

To rapidly detect new pathogens and vectors for disease or chemical (food borne) threats, we require a sample group to act as "bio-defenders" of the general population. These defenders will protect the nation by individually reporting their own health and submitting anonymized early warning health metrics. Ideally this group should contain people most likely to be exposed early on to new threats. We propose that airline crews, public transportation workers, postal workers, military service personnel, and other public service workers would provide a representative group. They are exposed to large numbers of people daily. In the case of airline crews, their location, schedules, and travel history already known in a flight database, as is the flight schedule of passenger (in case we later discover a contagious pathogen was present on a specific flight). The bio-defenders will be a first line of defense for the general population. They will not require constant medical attention, but will self-monitor and self-report through a web based information system, acting as a data source for the CHS detection subsystem.

3.1.1. *Data definition*

What data is required to detect biochemical threats before clinical evidence emerges? To develop the required data paradigm we analyzed the CDC database of infectious diseases [9]. The analysis focused in particular on diseases with relatively short incubation times and diseases most likely spread to travelers. Some of these diseases are also considered candidates for weaponization. We studied the symptoms of 45 infectious diseases and families of diseases (see Table 1). For each of these diseases, we correlated the expression of 38 common symptoms over the 45 types of pathogens (Table 2). The 38 symptoms (or symptom groups) are ranked by how often the symptom is reported. Common symptoms were given a weight of 1. Where a symptom is listed as "sometimes" expressed, it was given a weight of 50%. Note that this analysis is only valid for populations, as individual symptoms will vary. Therefore any sample group to be used for early detection of disease must contain a statistically significant population. This is discussed in greater detail below.

Table 1. Diseases analyzed.

African trypanosomiasis	Malaria
Amebiasis	Measles
Campylobacteriosis	Mumps
Chagas' disease	Rubella
Cholera	Meningitis
Coccidioidomycosis	Noroviruses
Cyclospora	Onchocerciasis
Dengue fever	Plague
Diphtheria	Rabies
Ebola Hemorrhagic Fever	Rickettsial Infections/Typhus fevers
Epstein-Barr virus	Spotted fevers
Tetanus	Rotavirus
Pertussis	Salmonella
E-Coli	SARS
Japanese Encephalitis	Scabies
Filariasis, Lymphatic	Shigellosis
Giardiasis	Smallpox
Hepatitis A(BCE)	Tuberculosis
Histoplasmosis	Typhoid fever
Influenza	Varicella (Chicken Pox)
Leishmaniasis	Vibrio parahaemolyticus
	Viral hemorrhagic fevers
	West Nile Virus
	Yellow fever

Of all 38 symptoms or classes of symptoms analyzed, "fever" is by far the most common symptom. It is easy to measure and is manifest in over 60% of cases. Many of the conditions that report fever also report one of the following three common symptoms listed in Table 2. If you remove all conditions that show fever, the next two most common symptoms for diseases that to not exhibit fever are "rash/spots/or Bleeding under the skin", followed closely by "diarrhea". Together, these

Table 2

Symptom	Percent Exhibiting
Fever	**63%**
Fatigue/lethargy/malaise/sleepiness	37%
Headaches	37%
Abdominal cramps/ stomach pain	33%
Rash/spots/bleeding under skin	**32%**
Diarrhea	**31%**
Nausea	27%
Vomiting	27%
Muscle aches/myalgia	26%
Joint pain	17%
Loss of appetite/anorexia/weight loss	17%
Cough	16%
Sore throat	16%
Swollen lymph nodes	14%
Chills	13%
Swelling	**13%**
Diarrhea (bloody)	11%
Disorientation,delirium,dizziness	9%
Jaundice	7%
Spasms/convulsions	**7%**
Stiffness (neck or abdomen)	7%
Chest pain	4%
Difficulty swallowing	**4%**

cover 12 of the remaining 16 conditions. The next 9 symptoms are often reported with those two. To detect the last 4 conditions (Tetanus, Pertussis, Lymphatic Filariasis, Rabies) one would need to detect "spasms/convulsions", "difficulty swallowing", and "swelling" in that order. At least one of these types of symptoms would be expressed by any of the 45 types of infection.

There is no reason to restrict the early warning data to the 6 spanning symptoms described above. The important point is that with the exception of fever, all of the first 23 symptoms or "complaints" could be reported by the subjects before seeing a medical professional. Since almost all of these symptoms can easily be reported by individuals in the sample group the CHS detection set could, in fact, include the first 23 data types, requiring only a digital thermometer to measure fever.

3.1.2. *Mechanics of data collection*

We propose to use a web based system to collect the reports from the bio-defenders. The web presents an ideal and cost effective infrastructure for such a data collection system. The symptoms covered are easily detectable by a layman and do not require the bio-defender to see a medical professional. The data will be collected in a secured manner and communicated into a centralized database. Fever or temperature will be measured by a simple digital thermometer or paper thermometer test strip.

The IT infrastructure required for a prototype CHS system will be based on existing state of the art technology including a database server integrated with a web server using JDBC. The bio-defender will enter data using web form. This data exchange will make use of existing health standards and will be communicate as HL7 xml messages and documents to the web server where the reported symptoms, time, location, id, etc will be entered into a relational database[a].

Figure 1: CHS data entry.

---

[a] The IBM solution CGv2 (recently released by the Life Sciences Organization) consists of such a relational data base designed around the HL7 Clinical Document Architecture.

3.1.3. *Incentives*

Bio-defenders participating in CHS are chosen from an "at-risk" group. An at-risk group is, by definition, one likely to be exposed early on to new pathogens. As discussed in detail in Section 3.4, participation in CHS gives the bio-defenders early access to medical care and early warning to emerging threats. This early access to information and care makes participation safe and attractive. Bio-defenders would receive treatment not only when they themselves exhibit symptoms, but also if and when their peers (others with whom they make frequent contact) exhibit significant symptoms. In the airline pilot program, the responders could also be nurses and medical staff present at large airports, or local hospital and urgent care centers. These professionals would call on medical experts at CDC and infectious disease experts as needed to diagnose any outbreaks and establish general reaction protocols to each incident per existing policy.

There are several incentives for the sample group to act as bio-defenders for the general population.

- Patriotism. Many bio-defenders will volunteer.
- Advantageous to Individual. The bio-defenders are chosen from a "high risk" group. The risks of exposure to these groups are real. Participating in CHS as a bio-defender also gives the bio-defender early access to medical care and early warning to emerging threats. They are safer participating than not participating.
- Bio-defenders' peers would receive treatment as well.
- Privacy Protection. The bio-defender's data will be anonymized for the analysis in compliance with HIPAA[b] regulations. Threat detection does not require revealing the identity of individual bio-defenders. Personally Identifying Information is only required or used in order to provide the bio-defenders with personal care.

---

[b]HIPAA – Health Insurance Portability and Accountability Act.

Bio-defenders are not "canaries". The bio-defenders are patriotic volunteers who would receive rapid access to care in return for their effort. Furthermore, in the event of a real biochemical attack the traditional medical response is essential to providing a national response plan. CHS does not replace physicians or the CDC – it merely provides them with a timelier alert to emerging threats.

To expand CHS in the future to provide an even finder data grid, one can consider wider groups of bio-defenders and other data collection paradigms. The barriers to these alternatives are sociological not technological. We propose to work with airline crews as part of a proof of principle for a broader CHS system. As part of the CHS effort, we propose consultation with community groups to gather other ideas and engage in surveys to determine which scenarios make the most sense as part of a broader CHS system.

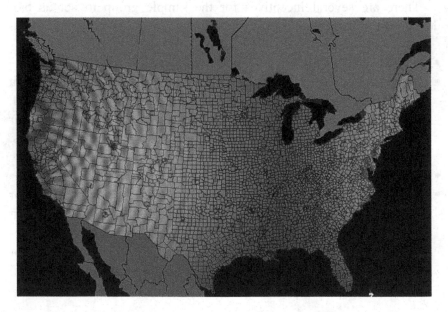

Figure 2: This is an artistic representation of a "model epidemic" wave with two nucleation sites in California. The example shows a propagating periodic wave with a limit cycle. The map shows areas within the US with country lines drawn.

3.2. *Creating a national health map*

Web forms entered by individual bio-defenders will be continually analyzed based on location, time, symptoms, etc. to generate a real time "health map" of the United States with a resolution based on U.S. counties. In order to integrate the map with epidemiological models, forecasts, data mining algorithms, etc., the individual data records will be analyzed to generate time dependant statistical data, by county, for the total sample numbers, number of individuals expressing symptoms (infected by some pathogen), number of individual recovering from infection, and individuals likely to have been exposed (once an infectious condition is detected). The health map will track not only susceptible, exposed, infected, and recovering populations but also the relevant time derivatives. It will make use of the bio-defenders known location and travel patterns to estimate exposure to infection and corresponding change in exposure rates.

In direct analogy to weather forecast, the continuously time varying health of our nation is a physical example of a "dynamical system" [29–34]. One can describe a dynamical system by a deterministic mathematical function that evolves the state of the system forward in time [35]. Different diseases, arising with different initial conditions, exhibit unique patterns in time and space. For example, for some natural illness (e.g., the flu) one can observe periodic expression of illness (limit cycles) decaying to background levels of steady state health [29]. Dynamical systems also exhibit non-linear and transient affects. Since the variables that describe a dynamical system are coupled, the dynamics can be studied by monitoring a reduced set of variables even thought the system as a whole may be quite complex and evolve in a high dimensionality phase space. This framework provides a basis for detecting non-linear affects (e.g., bio-terrorism) and for detecting unusual or previously unknown patterns of infection.

3.3. *Detection*

CHS is optimized to *detect* the onset of new illnesses – *not* to diagnose them. To accomplish this, the system will process the health map data (symptom and combination of symptoms), using two independent approaches, one based on epidemiological modeling and one based on data mining of spatio-temporal patterns.

3.3.1. *Modeling*

The national health map can make use of state of the art epidemiological models and improve our ability to create new spatial epidemiological models. Together with a baseline of time varying health data these new models will allow us to move beyond early warning and towards forecast of naturally occurring epidemics. Better spatial models can also be advantageous in planning immunization strategies. Consider, for example, the standard model for the time varying fraction of Susceptible, Exposed, Infectious, and Recovered (SEIR) of individuals within a population (Fig. 3). The SEIR model treats the entire population as if all individuals interacting "within a stirred beaker". Susceptible individuals become exposed by infectious individuals with some constant rate [29–36, 39]. There is no spatial component to this standard model. Enhanced SEIR models divide the population into risk groups but do not account for geographic variations. J. Epstein *et al.* in "Toward a containment strategy for smallpox bio-terror: Towards a containment approach" [40] show the importance of considering spatial variables in modeling the spread of smallpox. The "immunological landscape" has important affects on the spread of the virus. This is clearly evident in their individual based model even though it includes only two neighboring populations. In a seminal experimental study, Cummings *et al.* demonstrated the importance of spatio-temporal effect in their experimental data that demonstrates "Traveling waves in the occurrence of dengue hemorrhagic fever in Thailand" [41].

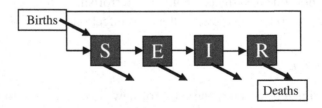

Figure 3: The standard SEIR model.

An ideal spatial model would merge a proven epidemiological model with spatial-temporal health data based on an accurate model of traffic flow. [42] For example, consider the following modification to the SEIR model of epidemics spreading between U.S. counties. In the standard model the rate at which Susceptible individuals are exposed to infection by infectious individuals is considered fixed by a (constant) rate of transmission, b. In a spatial model, this term would be preserved as a local or "on site" transmission probability. However, communication or transmission of disease from one site to another would involve a product of two probabilities, namely the rate of transmission times the probability of moving from the neighboring site to the site in question.

In a practical model it is necessary to define a smallest space. The smallest space is defined by practical data gathering constraints, and sets the resolution of both the transport model and the epidemiological model. For example, we may define a "space" as (e.g.) a county. A country is made up of many counties (Fig. 2). A county has a population, and may have one or more neighbors. A county object may also contain other important spatial variables such as a number (from zero to many) hospitals or medical centers, airports, roads, interstate highways crossing edges with neighboring counties, etc.

For *each* population (by county) define:

- m as the per capita death rate (assumed equal to the birth rate in the SEIR model).

For *each* disease define:

- b as the transmission coefficient measured as individuals infected per infected individual per unit time.
- a-1 as the average latency period.
- g-1 as the average infectious period.
- r-1 as the average period of immunity.

The spatial model adds a point to point (county to county) coupling parameter, ρ_t. We define ρ_t as the rate at which people from one county visit a connected county. The value of ρ_t may vary for each county-

county connection depending on the presence or absence of airports, interstate highways, length of border, local population density, policy (temporary action to isolate an infected population), etc.

In a modified SEIR model, the rate equations for the Susceptible and Exposed populations become (for each county and for each disease)

$$dS(t)/dt = (1-m)S(t) - bS(t)I(t)/P - \Sigma_n b\rho_t S(t)I_n(t)/P + rR(t)$$

$$dE(t)/dt = bS(t)I(t)/P + \Sigma_n b\rho_t S(t)I_n(t)/P - (a+m)E(t)$$

where Σ_n is a sum over connected neighbors. $S(t)$ is the susceptible population at the county in question and $I_n(t)$ is the infected population in the connected county. This new sum term introduces the spatial dependencies. The following two equations remain unchanged from the non spatial model.

$$dI(t)/dt = aE(t) - (g+m)I(t)$$

$$dR(t)/dt = gI(t) - (r+m)R(t)$$

In this example, the parameters b, a, g, and r, are disease specific parameters in the SEIR epidemic model. The detailed geographic data, along with local population densities and the pair-wise transit rates ρ_t define a traffic flow model for the population.

Creating of an CHS real time weather map with sufficient resolution or detection network would allow the development and validation of an accurate spatial epidemic model in direct analogy to weather modeling. Creation of new models that have real time predictive capabilities based on time dependent national health data can be used in planning immunization strategies in response to epidemics of the future. An epidemiological model can be applied to (filtered) health-map data to generate a set of estimated model parameters using nonlinear regression analysis in direct analogy to the methodologies used to develop weather models (Fig. 4).

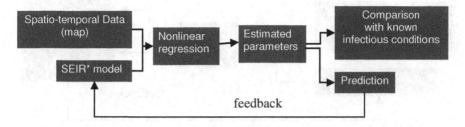

Figure 4: Modeling the spatio-temporal patterns of disease.

The estimated parameters may then be used to:

- Project forward the evolution of a particular pattern of infection
- Compare the parameters for a time evolving infection to known historical data (by type of infection)
- Feed back. Compare "today's" data with "yesterday's" estimate to refine the model

3.3.2. *Data mining*

Epidemiological models provide a well tested approach towards predicting the time evolution of most natural illnesses. However, when dealing with unknown or newly emerging infectious diseases [41], there is always the possibility that a specific model will leave out essential processes needed to predict the time evolution. In such cases it is important to also enable model independent analyses. The CHS health-map will provide a spatio-temporal correlation tool to compare patterns emerging in the health map data with historical patterns of known diseases. It will also provide general purpose "data mining functions, an interface so users can study correlation with their own data, and an ability to evaluate the CHS health map data with other tools such as the public SatScan package developed by Kulldorff *et al.* This package was developed specifically to perform statistical analysis on spatio-temporal data for detection of disease clusters, recognition of geographically significant patterns of infection, etc. [43]

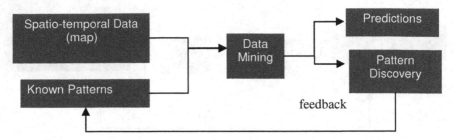

Figure 5: A model independent data mining analysis will be used to discover new spatial pathways.

In the simplest pattern matching scenario, a subscriber to the CHS health map may define a filter or query pattern that if expressed by the map data would trigger an alert. A simple pattern, for example, might set an alert for the near simultaneous expression of a set of symptoms at airports. Such and alert might but designed to detect acts of bioterrorism at airports. This tool will also provide a general purpose interface to allow mapping of user defined data types and correlate these new data patterns against current health data. This will allow users to test new theories and models.

As a concrete example, consider an outbreak of a new variant of Avian-Influenza. In this hypothetical case suppose the new virus is transmitted to humans via mosquitoes from a particular species of migratory birds. A spatial SEIR model based on a human traffic flow network would not anticipate or predict observed human infections originating from the migratory path specific to birds carrying the virus. A general health map interface allowing researcher to submit, over the web, spatio-temporal data corresponding to the observed bird migration patterns (and/or mosquito populations varying by rainfall) would allow quick evaluation of new hypothetical vectors of infection. In another example, the "EpiSpire" project [30] demonstrated that the spatial distribution of Hantavirus infections could be accurately predicted by satellite imagery of vegetation die-off (which in turn triggers migration of mice into human habitats as the mice look for food). A general tool to discover such complex pathways will be valuable in understanding emerging infectious disease and in developing new predictive models.

3.4. *Reaction*

The CHS system will react according to two separate protocols whenever an infectious disease is detected. The first protocol focuses on the health of the individual bio-defenders contributing electronic health data to the system. The second protocol is a response to detected patterns of infections.

3.4.1. *Reaction in support of affected individuals*

As part of the incentive for bio-defenders, they will receive treatment as soon as they report significant symptoms. As added incentive, the CHS system knows not only about individual bio-defenders that report illness, it also detects individuals that may have been exposed to dangerous infections even if they are asymptomatic themselves. Both groups of people will be directed to seek individual treatment. CHS will support this individual treatment by providing the specific electronic health reports for affected or exposed individuals to the participating medical team. These reports will make use of current medical information standards as discussed in Section 4.

3.4.2. *Communication and feedback to existing centers*

The CHS system does not replace programs optimized to diagnose the cause of new diseases. Rather, it is synergistic with these programs by providing a running real time national health map. CHS is complementary to existing efforts including programs such as PHIN [17], PulseNet [14], FERN [19], LRN [18], SensorNet, and others. The bio-defender data acquired by CHS augments the data collected by "Essence" [26], RODS [28], and BioSense [9]. The regional and national Center for Disease Control CDC offices, the Department of Homeland Security (DHLS), the Terrorist Threat Integration Center (TTIC), Hospital Emergency Rooms, and regional infectious disease centers can all make use of the CHS health map.

CHS does not provide specific diagnostic data or treatment recommendations. It is intended to provide early warning and better information to other agencies charged with diagnosing emerging infections and discriminating potential sources or epidemics from natural illnesses such as the common cold and yearly influenza cycles. The bio-defenders, like the general population, will experience common illnesses. By constantly mining the bio-defender sample data on an ongoing basis, CHS will establish a national baseline and detect the patterns, changes, and correlations in time and space to discover statistically significant new events. Such events will reported to the appropriate centers using current medical information standards as discussed in Section 4.

The CHS health map web interface makes information available to any and all national agencies, senor-net monitors, hospitals, and treatment centers. Subscribing centers will also have available a java CHS health map application allowing them to interact with the data and build their own models. The data will be exposed using standards as discussed in Section 4. In this way CHS complements data currently exchanged between health agencies following the formal diagnoses of emerging illnesses.

3.5. *Cost considerations*

CHS is designed to achieve its goal as an early warning system in a cost affective way.

- CHS makes use of standard internet technologies to transmit and process data. No major new IT infrastructure is required by subscribers to the system. No new technologies are required to store or serve the health map data. By leveraging existing state of the art technology the CHS core infrastructure is efficient and cost affective.
- Since the bio-defenders themselves provide the data source, CHS puts no added burden on already overworked healthcare providers. When a new threat is detected, the existing systems respond per current protocols.
- By providing early warning of new patterns of infections, CHS should actually lower the overall medical cost of response. Early warning allows existing agencies to react faster in developing and communicating immunization strategies, release and dispatching medication, and containing new diseases. If fewer people become ill, the cost of treatment and burden on the medical caregivers is reduced.

4. Interaction with the Health Information Technology (HCIT) World

To realize the advantages in a system such as the CHS system, information must freely flow between the system and the various organizations and agencies that deal with public health and health care. In recent years there is a move towards a paperless information exchange in the health care realm. Standards are being developed to facilitate this information exchange. In the following paragraphs we will describe how the CHS system will leverage standard-based interfaces to communicate with other HCIT systems.

Key to the CHS system is the periodic reports of the bio-defenders. In addition to the immediate use of the bio-defenders reports in early detection of pandemic events, these reports may be used in other ways to improve the response of the medical system as a whole to such events. To allow data exchange between our system and the HCIT world, the reports should be maintained in the individual's Electronic Health Record (EHR) according to the appropriate standards and specifications. HL7[c] is the dominant organization which develops such standards and in the CHS system we will use various specifications depending on the content and purpose of the data as follows: For the information exchange regarding the individual bio-defender care we will use HL7 V2/V3 messages such as the Orders & Observations messages that convey lab test orders along with its results or the Pharmacy messages that deal with prescribing, dispensing & administration of medications. HL7 V2 message are widely used by healthcare providers in the US and thus are good basis for the communication between CHS and the HCIT world. Nevertheless, for clinical documentation we will use a newly created standard from HL7 called CDA (Clinical Document Architecture) which is capable of holding clinical narrative as well as structured data in the form of coded observations, procedures, medications and so forth. The CDA standard will be used for example to refer bio-defenders to healthcare providers

---

[c] An ANSI-Accredited Standards Developing Organization. See appendix A for more information on HL7 as well as on IBM engagement and contributions to HL7.

by having a standardized referral letter sent to the provider, resulting at better and more efficient care.

For the information exchange regarding the public health, we will be using the new HL7 V3 family of standards and especially the messages defined in the Public Health chapter. Based on the same data structures that were used to initially populate the CDA document for the individual care, it is possible to create messages that agencies like the CDC are capable of accepting, parsing and acting accordingly. We will design a new message which we call - NNBT (Nationally Notifiable Bio Threat), derived from the HL7 V3 Public Health Domain Information Model. This new message will resemble in its structure to the existing NND (Nationally Notifiable Disease) message. It is derived from the Public Health HL7 Domain Model and thus based on the HL7 RIM. Deriving the message in this way assures semantic interoperability across all agencies that deal with public health and base their system interfaces on HL7 V3 (like the CDC).

If and when an actual medical necessity emerges, the medical professionals providing care to the bio-defenders will be able to query CHS in real time for the patients' records. This data may be transferred to their own system in a seamless error free manner for the benefit of the patients. These records will not only contain the latest report from the bio-defender but the whole longitudinal[d] information regarding that individual's health condition accumulated in our system, providing valuable details to the care giver.

Data exchange between CHS and the care givers' system may also occur in the opposite direction. When a bio-defender seeks medical attention in relation to an event detected by our system, the care-giver's diagnosis and treatment information may be propagated to our system. This information (after being anonymized) may then be shared with care-givers of other bio-defenders exhibiting similar symptoms.

---

[d] Note that the ABDN is not designed to hold longitudinal EHRs which are envisioned to be birth-to-death entities with comprehensive health data regarding an individual, nor it is designed to function as a complete EHR system of a healthcare provider. Rather, the ABDN system is persisting medical records in standardized formats (like HL7) that could be exchanged with standard-compliant EHR systems.

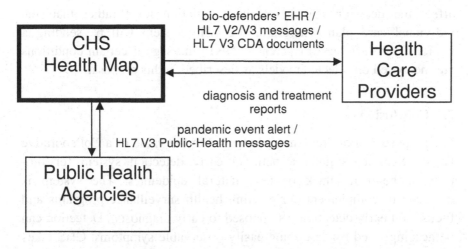

Figure 6: Standard-based information exchange between the CHS system and the other health organizations. The figure depicts the main interfaces between the CHS system and the HCIT world. For each interface appropriate messaging standards are specified.

In the longer run, the diagnosis, treatment and prognosis reports may be used as inputs to learning algorithms within our system. These learning algorithms will enable continual improvements in the system's accuracy in detecting infectious disease and forecasting the spatio-temporal evolution of disease. Furthermore, they may even be used to build an intelligent system able to help track causes of new epidemics.

As aforementioned, the CHS system is propagating its data in two parallel paths -the individual and the population paths.

It is important to note that while a CDA document should contain one and only one patient[e], the public health messages could aggregate several cases and report them in one message where the same context applies to all subjects. Those subjects could be either identified or anonymous. In addition, a CDA document is supposed to be attested and signed by a legal authenticator (depending on local regulations) with all parties fully identified with unique identifiers. In contrast, population-oriented messages are more about the real-time and aggregated data that need to

---

[e]With the rare exception of a family therapy encounter.

affect the decision making at the population level rather than the individual care plan. Therefore, the CHS system will be sending a Nationally Notifiable Bio Threat (NNBT) message if certain conditions are met based on selected models as described in this proposal.

5. Conclusion

We propose a pilot program to demonstrate the value of a Collaborative Health Sentinel system, a national early detection system for bio-terrorist/chemical attack or new natural epidemics. The system is designed to complement our existing health surveillance programs and focuses on early detection (as opposed to early diagnosis). Detection can be accomplished by describing easily observable symptoms. CHS takes advantage of existing information technologies to provide a cost effective web based data collection and publishing system. Data entry does not require the participation of medical professionals and so does not add new burdens to overloaded healthcare workers.

Early detection does require an ongoing *sampling* of the health state of our populace. To accomplish this we propose a pilot study with a core team of "bio-defenders". The bio-defenders are chosen from groups likely to be exposed early on to new infectious disease. We also provide incentives for the bio-defenders to participate. As participants they receive early warning of new health threats and benefit from rapid, even proactive, treatment.

The real time data available through the CHS health map is a service that any interested health organization can subscribe to and access over the web. Integrating the health map with modeling tools, data mining tools, and pattern recognition technologies allows users to go beyond observation and to detect newly emerging diseases. These tools will allow researchers to make predictions, evaluate and enhance health forecast models, and study the correlation between emerging patterns of infection with postulates for unusual vectors of infection.

The CHS health map will also be a resource to those agencies and medical professionals charged with reacting to and containing potential epidemics. As an early warning system, the CHS health map makes it possible for physicians to quickly ascertain if unusual symptoms reported by an individual represent a larger pattern of infection or an isolated incident. CHS complements the effort of CDC and other agencies in correlating real time health reports with the known locations and travel patterns of bio-defenders. When a new pattern of infection does occur, CHS provides communication and feedback to existing centers based on current standards and messaging protocols. This gives traditional agencies timely access to the EHR of affected bio-defenders as well as a real time big picture view of possibly nation wide threats. Such information is useful both in diagnosing the cause of disease, finding points of origin, and in planning containment and immunization strategies.

References

[1] Renton, S., "The Bioterrorism act of 2002, Protecting the Safety of Food and Drug Supply", International Trade and Customs News (Sept. 2003), http://www.gtlaw.com/practices/intl_trade/newsletter/200309.pdf

[2] Kuliasha, M. A., "Challenges of International Trade: Balancing Security and Commerce", http://www.ornl.gov/~webworks/cppr/y2003/pres/116647.pdf

[3] Heinrich, J., "Improvements to Public Health Capacity are Needed for Responding to Bioterrorism and Emerging Infectious Diseases", USGAO (May 7, 2003) http://www.gao.gov/new.items/d03769t.pdf

[4] http://en.wikipedia.org/wiki/2001_anthrax_attack

[5] Joshua Lederberg report to Committee on Foreign Relations (Aug. 24, 2001) *as published in* Emerging Infectious Diseases 10, 7 (July 2004)

[6] http://www.rppi.org/wtc/anthrax.html

[7] Tom Ridge, April 28, 2004, http://www.defenselink.mil/transcripts/2004/tr20040428-depsecdef1383.html

[8] see: Case Definitions for Public Health Surveillance (MMWR 1990;39[No. RR-13]) http://www.cdc.gov/hiv/pubs/mmwr.htm

[9] Center for Disease Control and Prevention (CDC), http://www.cdc.gov/

[10] Health Alert Network http://www.phppo.cdc.gov/han/index.asp

[11] National Electronic Disease Surveillance System http://www.cdc.gov/nedss/

[12] Epidemic Information Exchange http://www.cdc.gov/epix/

[13] Health Level 7 Reference Information Model http://www.hl7.org/library/data-model/RIM/modelpage_mem.htm

[14] PulseNet national food borne disease surveillance network http://www.cdc.gov/pulsenet/

[15] Epidemic Intelligence Service http://www.cdc.gov/eis/

[16] The Emerging Infectious Disease Journal
      http://www.cdc.gov/ncidod/EID/index.htm
[17] Public Health Information Network http://www.cdc.gov/phin/
[18] The Laboratory Response Network http://www.phppo.cdc.gov/nltn/pdf/LRN99.pdf
[19] The Food Emergency Response Network
      http://www.crcpd.org/Homeland_Security/Food_Emergency_Response_Network.pdf
[20] The Electronic Laboratory Exchange Network
      https://www.elexnet.com/elex/index.jsp
[21] Anick Jesdanun, Associated Press (June 26, 2004)
      http://www.startribune.com/stories/484/4847829.html
[22] The National Animal Health Laboratory Network
      http://www.csrees.usda.gov/nea/ag_biosecurity/in_focus/apb_if_healthlab.html
[23] Enserink, M., Science 301 294-296 (July 18, 2003), and Enserink, M., Science 301
      299 (July 18, 2003), and Normile D., Enserink, M., Science 301 297-299 (July 18,
      2003)
[24] http://www.cdc.gov/ncidod/dvbid/westnile/
[25] http://www.cdc.gov/ncidod/dvbid/dengue/
[26] See for example, *Syndromatic Surveillance Conference* http://www.syndromic.org/
      and http://www.syndromic.org/con2.html, *and* Hurt-Mullen, K. *et al.*, "Local
      health department applications of essence biosurveillance system"
      http://apha.confex.com/apha/132am/techprogram/paper_93642.htm
[27] Hoffman, M.A. *et al.*, "Multijurisdictional Approach to Biosurveillance,
      Kansas City", Emerging Infectious Diseases, V. 9, No 10, Oct 2003.,
      http://www.cdc.gov/ncidod/EID/vol9no10/03-0060.htm, and (Essence I & II)
      http://www.geis.fhp.osd.mil/
[28] Tsui, F. C., J. U. Espino, *et al.* (2003). "Technical Description of RODS: A
      Real-time Public Health Surveillance System." J Am Med Inform Assoc.
      www.jamia.org, and http://openrods.sourceforge.net/index.php?page=publications
[29] See for example http://bill.srnr.arizona.edu/classes/195b/195b.epmodels.htm and
      references therein.
[30] C.-S. Li, C. Aggrarwal, M. Campbell, Y.-C. Chang and G. Glass, V. Iyengar,
      M. Joshi, C.-Y. Lin, M. Naphade, J. R. Smith, B. L. Tseng, M. Wang, K.-L. Wu,
      P. S. Yu, "EPI-SPIRE: A Bio-surveillance System for Environmental and Public
      Health Activity Monitoring," *Proc. IEEE Intl. Conf. on Multimedia and Expo
      (ICME)*, Baltimore, MD, July, 2003. and
      http://www.research.ibm.com/networked_data_systems/esip/
[31] Liu, W-M., Hethcote, H. W. and S. A. Levin. 1987. Dynamical behavior of
      epidemiological models with nonlinear incidence rates. *J. Math. Biol.* 25: 359-380.
[32] Olsen, L. F. and W. M. Schaffer. 1990. Chaos *vs.* noisy periodicity: Alternative
      hypotheses for childhood epidemics. *Science.* 249:499-504. Schaffer, W. M. 1985.
      Can nonlinear dynamics elucidate mechanisms in ecology and epidemiology? *IMA
      J. Math. Appl. Med. Biol.* 2: 221

[33] Schaffer, W. M. and L. F. Olsen. 1990. Chaos in childhood epidemics. Pp. 187-190. **In**, Abraham, N B., Albano, A. M., Passamante, A. and P. E. Rapp. *Measures of Complexity and Chaos.* Plenum Press.

[34] Schaffer, W. M., Olsen, L. F., Truty, G. L. and S. L. Fulmer. 1990. The case for chaos in childhood epidemics. Pp. 139-167. **In**, Krasner, S. (ed.). *The Ubiquity of Chaos.* AAAS. Washington, D.C.

[35] see for example http://en.wikipedia.org/wiki/Dynamical_system

[36] Anderson, R. M. 1982. *Population Dynamics of Infectious Diseases: Theory and Applications.* Chapman and Hall, New York.

[37] Anderson, R. M. and R. M. May. 1982. Directly transmitted infectious diseases. *Science.* **215**: 1053

[38] Bartlett, M. S. 1960. Stochastic Population Models in Ecology and Epidemiology. Methuen, London.

[39] Bailey, N. T. J. The Mathematical Theory of Infectious Diseases and its Applications. Griffin, London.

[40] Epstein, J., *et al.*, "Toward a Containment Strategy for Smallpox Bioterror: An Individual-Based Computational Approach", CSED Working Paper 31 (Dec 2002) http://www.brookings.edu/dybdocroot/es/dynamics/papers/bioterrorism.htm

[41] Cummings, D.A.T., *et al.*, "Traveling waves in the occurrence of dengue hemorrhagic fever in Thailand", Nature V 427 p344 (Jan 22, 4004).

[42] Haberman, R., Mathematical Models: Mechanical Vibrations, Population Dynamics, & Traffic Flow. Soc for Industrial & Applied Math (Feb 1998).

[43] Kulldorff M, *et al.* C. Evaluating cluster alarms: A space-time scan statistic and brain cancer in Los Alamos. American Journal of Public Health, 88:1377-1380, 1998. and http://www.satscan.org

Appendix A – HL7

Information exchange in the healthcare arena should be done based on internationally-recognized medical informatics standards to assure semantic interoperability of disparate health IT system and better consistency across the various medical domains[f]. HL7 (an ANSI-Accredited Standard Developing Organization[g]) is the dominant SDO for healthcare information in the USA along with more than 25 international affiliates around the globe. HL7 has started developing standards for healthcare in 1987 and traditionally it has been focusing on ADT (Admission-Discharge-Transfer) messages in hospitals. HL7 V2 specifications are widely implemented in hospitals throughout the USA and are supported by every major medical informatics system vendor in the US.

They cover domains like ADT, orders and observations, pharmacy and medical records management. In the recent years, HL7 has published a number of specifications that were derived from the new V3 RIM[h] (Reference Information Model) – an innovative component of the new V3 version. Early adopters are coping with V3 implementations at this time and it's mostly notable in national projects such as the multi-billion UK NHS NPfIT (New Program for IT) as well as in public health and decision support projects such as those where the CDC and the FDA are involved in the USA. Both agencies have chosen HL7 V3 as their strategic standard of information exchange.

IBM is a benefactor of HL7 and active organizational member. IBM representatives are attending the HL7 Working Group meetings and contribute to the standard developing process. In particular, IBM representatives are active in the following groups:

---

[f] See for example the NHII (National Health Information Infrastructure) web site at: http://aspe.hhs.gov/sp/NHII.

[g] See http://www.hl7.org.

[h] The HL7 RIM (Reference Information Model) is the core model from which any HL7 V3 specification is derived by methods of refinement only. There is no way to extend the model or customize it, rather merely refine it. In this way, any V3 specification that was created using the HL7 Development Framework (HDF) will have the core components like observations, procedures, medications (and their interrelationships) represented in the same way, thus laying the ground for true semantic interoperability.

The **EHR** Technical Committee which has recently published the EHR Functional Model DSTU (Draft Standard for Trial Use) which has been initiated by of the CMS agency of the USA government.

- The Structured Documents Technical Committee which is now completing the second release of the **CDA** (Clinical Document Architecture) specification, merging structured and unstructured clinical data based on the core HL7 V4 Reference Information Model.
- The **Clinical-Genomics** Special Interest Group which IBM has initiated two years ago, dealing with the innovative use of genomic data in actual healthcare practice.
- The RCRIM (**Clinical Trials**) Technical Committee which develops standards to represent all data exchange between the healthcare providers, the clinical research organizations and the FDA.
- The **Imaging Integration** Technical Committee which cope with the way medical imaging information (e.g., in DICOM format) is correlated with the clinical data and is represented in a consistent way with HL7.

CHAPTER 9

A MULTI-MODAL SYSTEM APPROACH FOR DRUG ABUSE RESEARCH AND TREATMENT EVALUATION: INFORMATION SYSTEMS NEEDS AND CHALLENGES

Mary-Lynn Brecht and Yih-Ing Hser

Integrated Substance Abuse Program, UCLA

Abstract

Drug abuse research and evaluation can benefit from the implementation of a multi-modal system including data integration and mining components for modeling complex dynamic phenomena from heterogeneous data sources. Such a system could help overcome current limitations of distributed responsibility for and location of relevant data sources, the heterogeneity of these sources, and the resulting difficulty in accessing and analyzing such disparate data sources in a timely manner. We present a users' perspective on selected topics relating to the initial stages of development of an online processing framework for data retrieval and mining to meet the needs in this field. We describe several challenges in designing and implementing such a system, primarily related to the real-world context and applicability of such a system for drug abuse research.

1. Introduction

There is currently a need and opportunity to expand the empirical modeling capabilities in the general area of substance abuse research across topics involving evaluation, policy development, law enforcement, and clinical practice as related to substance abuse prevention, interdiction, and treatment. Facilitating the use of heterogeneous data sources related to substance abuse and its context, consequences, prevention, and treatment through data integration and mining can enhance opportunities to model this complex phenomenon. In turn, results can improve the responsiveness and adaptability of analytic models to provide support for decision-making at client, program, and government agency levels. We discuss below the preliminary development of a multi-modal infrastructure to support data integration and mining, and several challenges we foresee in designing and implementing such a project, primarily from a system user perspective.

While current technology implementation is still uneven across constituent substance abuse research and evaluation communities and while most data systems are autonomous, several important building blocks are in place to support the development of integrated information systems. The need for data-based support for policy and planning, as well as many of the associated difficulties, have long been recognized in the field of substance abuse (e.g. Anglin *et al.*, 1993; Camp *et al.*, 1992; Hser, 1993). In addition to this historic context, the current context is providing motivation for applying new approaches to meeting these needs. The staggering economic and social costs of substance abuse provide an increasing imperative to improve prevention, treatment, and law enforcement strategies to ameliorate the problem, for example, estimated economic costs of illicit drug use at $181 billion in 2002, and 19.5 million people in 2003 with current illicit drug use (Office of National Drug Control Policy, 2004; Substance Use and Mental Health Administration, 2004a). There is a current emphasis on the implementation of evidence-based practice and policy, a recognized need for integrated systems for managing treatment delivery and performance monitoring, and federal and international initiatives to address these needs (e.g. Center for Substance Abuse Prevention, 2002; Coalition for Evidence-Based Policy, 2003; Daleiden and Chorpita, 2005; Global Assessment Programme on Drug Abuse, 2003; Lamb *et al.*, 1998; Simpson, 2002).

The field currently relies primarily on autonomous data sources, physical data archives, and some on-line analysis systems for selected analyses of individual data sources. For example, the Substance Abuse and Mental Health Archive (available through the Institute for Political and Social Research [ICPSR] data archives) and the Bureau of Justice Statistics allow on-line analysis of many databases relevant to substance abuse research and evaluation. But such analyses are currently limited to individual databases, with little integration across resources.

Data-sharing initiatives are improving the accessibility of a range of databases and have resulted in the availability to physical archives of more individual data sources and the possible physical integration of selected research data with homogeneous structure across a set of related studies (e.g. the Clinical Trials Network, sponsored by National Institute on Drug Abuse [NIDA]). The linking (physical integration) of data records from several sources (for example, to consolidate information on individual clients across drug treatment, mental health, and criminal justice systems) has proven useful for program development and resource acquisition, improving clinical practice, and providing data for policy development and assessment (e.g. Grella *et al.*, 2005; Hser *et al.*, in press; Jindal and Winstead, 2002). The (federal) Substance Abuse and Mental Health Administration is currently providing support for state infra-structure development to implement the National Outcomes Measurement System (NOMS), to enhance the accessibility and use of a set of standard data content useful for measuring prevention and treatment outcomes (Substance Abuse and Mental Health Administration, 2005).

Currently, optimal planning and near real-time adaptation of the component systems is often hampered by difficulties in integrating data, including the distributed responsibility for and location of relevant data sources and resulting difficulty in accessing these data by the diverse research, evaluation, policy, criminal justice, and clinical practice communities. These difficulties in integration also restrict the comprehensiveness of analytic strategies. These problems are further compounded by the complexity of analyzing such disparate data sources to provide timely empirical support for both immediate resource allocation and planning for future years, as well as comprehensive information and knowledge generation.

2. Context

The specific context within which we consider data integration and mining in substance abuse research is one of heterogeneous data sources and complex questions of interest to clinicians, researchers, and policy makers. Substance abuse is recognized as a chronic illness, with associated need for prevention and ongoing disease management, including integration of services and continuity of care (Leshner, 1999). For example, many substance users also suffer from co-occurring mental disorders, physical health problems, have contact with the criminal justice system through arrests, jail, prison, or because of parental status are involved with the child welfare system.

2.1. *Data sources*

In the general field of substance abuse research, we have a loosely affiliated diversity of players from many different governmental agencies and private entities and their data from local, state, and federal levels in areas of law enforcement (and criminal justice, in general), health, mental health, social services (welfare, child protective services, education), substance abuse treatment and prevention, departments of motor vehicles (DMV), liquor licensing and sales, research and evaluation.

Data from the diverse components comprise heterogeneous schema and structures, with different purposes, sources (e.g. self-report questionnaires or interviews, clinical assessments, administrative records, etc.), scale, and management systems. For example, client-level data and/or episode and event data that can be matched to clients, include medical records (including biological tests and MRIs), genetic profiles, substance use treatment records, interactions with mental health or social service agencies, DMV information on DUI [driving-under-the-influence] and/or accidents, arrests, incarcerations. [See Hser *et al.*, (2003) for discussion of current approaches to accessing and linking multiple data sources to evaluate treatment outcomes.] Services data provide another perspective, for example at the provider or program level including staffing, types of services, client mix, and costs. Surveillance data (prevalence and location) include arrests, drug-related emergency room episodes and deaths, drug seizures, purchases of components for drug manufacture, drug lab closures, satellite imaging (e.g. to detect drug agriculture), public health surveillance of problems and epidemics related to drug use (e.g., risk behaviors, HIV, hepatitis, sexually transmitted diseases), and internet transactions. Data describing research results provide still another perspective, including relationships among variables and effect sizes.

We expect that new data modalities will increasingly become available, and that data access and mining systems must enable their inclusion. For example, these might include wider availability of brain imaging data, genetic mapping, and spatio-temporal data, which many of our current substance abuse data retrieval and analysis systems cannot yet efficiently handle.

Data quality is a challenge for data mining and information integration in general and for the field of substance abuse in particular where the diversity of databases also represent a diversity of quality in terms of, for example, completeness of coverage, consistency of coding, and reliability (Chung and Wooley, 2003; Global Assessment Programme, 2003; Hser *et al.*, 1993; Seifert, 2005). Major federal initiatives from the Substance Abuse and Mental Health Administration (SAMHSA) have been targeting issues of data quality in order to promote consistency in drug treatment and prevention data across a set of core concepts and measures (e.g. NOMS); these efforts include infrastructure-building and training designed to maximize completeness and reliability of data (Camp *et al.*, 1992; Del Boca and Noll, 2000; Hser *et al.*, in press). Adequate documentation of database characteristics (data "provenance") becomes especially critical in information integration efforts (Chung and Wooley, 2003).

2.2. *Examples of relevant questions*

Examples of questions/issues to which data mining of heterogeneous substance abuse data sources are appropriate include the following, as illustrated below in relation to substance abuse treatment. A similar range of issues can be addressed relating to e.g. prevention efforts or to law enforcement strategies and drug interdiction efforts. Potential users of any integrated information system range from, e.g., a clinician providing treatment to substance users to the White House Office of National Drug Policy, and every level in-between. A data integration and mining system would allow us access to a broader diversity of data resources, in a timely manner, in order to explore more complex empirical models of dynamic phenomena.

Example 2.1 *Prescribe treatment for substance abusers.* A primary area of interest is that of identifying optimally effective treatment prescriptions for substance-abusing client(s) according to their specific characteristics and needs. These characteristics (profile) could be specified by the system user (e.g. a researcher, treatment provider, or policy maker) or extracted from existing empirical data and matched to the models stored in the system in order to generate recommended treatment prescriptions; that is, given the system user's specification of desired type of outcome or pattern of outcomes (e.g. treatment completion, no re-arrest and/or treatment readmission within a specified period, etc.), the system would assess data through statistical modeling or data mining processes to suggest optimal treatment strategies (e.g. which modality, length of treatment, specific types of services or supervision, etc.) for the given client/contextual characteristics. The effectiveness of the treatment could be continuously monitored, so that the treatment prescription could be continuously adapted to the particular client(s). The updated results would also stored by the system to be used in the continuing process of model revision. For example, a model might suggest an optimal treatment modality given a specified set of client characteristics; as this client remains in treatment longer (receives specific services, remains drug-free, has additional contact with the criminal justice system, etc.), the model would be updated to produce new estimates of likely future outcomes conditional on the continuously updated performance data. Expanded access to diverse data sources would allow inclusion of a wider range of contextual characteristics and treatment outcomes.

Example 2.2 *Prescribe treatment or evaluate potential outcomes for specific policies.* For example, the Substance Abuse and Crime Prevention Act in California has allowed substance abuse treatment for many non-violent substance-using offenders who would otherwise have been incarcerated (e.g. Longshore *et al.*, 2005). The treatment system has had to respond quickly to this policy perturbation, without ready access to relevant data and potential patterns of service use and outcomes. Data mining of dynamic data sources could support rapid response planning to maximize treatment system response.

Example 2.3 *Identify optimal allocation of resources to achieve maximal treatment success.* This is a variation of the above two examples focusing specifically on costs, for example maximizing treatment success for a specific cohort of substance-using clients when budgets are fixed.

Example 2.4 *Assess the need for type, location, and number of treatment services.* For example, needs assessment could utilize data mining of law enforcement and other criminal justice surveillance data, as well as medical and other social data, to identify the location, timing, and trends of problem development and to use treatment evaluation results to match potential needs to relevant services.

3. Possible System Structure

Our UCLA (University of California, Los Angeles) and IBM team has envisioned a multi-layer system, adapting existing technology to meet the specific needs for data integration and mining in the field of substance abuse research. We have developed a skeletal system structure to include a data and information layer to enable federated access from heterogeneous data archives through a unified query interface and an online processing framework to integrate data retrieval, spatial and time-series data mining, and other statistical modeling to allow optimal simultaneous retrieval and statistical analysis of heterogeneous data sources. Note, that our discussion is from the perspective of the very early conceptual stages of a "work in progress"; thus we have not yet addressed many technical and other details that will arise during the development and implementation process. Our suggested approach offers the potential of addressing difficulties posed by heterogeneous data (for example, in physical location, scale, format, semantics, and ownership) and the dynamic and complex phenomenon of substance abuse and related modeling challenges. Many other approaches may also be applicable and can provide examples and strategies to support our efforts (see, e.g. Ellisman and Reilley, 2004; Grethe *et al.*, 2005; Hernandez and Blanquer, 2005; Lacroix and Critchlow, 2003).

We expect that conceptual and applications development of such a system can proceed directly because of previous development of related components that have been and are currently in use in other fields. Commercially available products can be adapted to provide federated access of heterogeneous data sources; for example, Multiple Abstraction Level Mining of time series (MALM) was developed to enable search and retrieval of time series pattern based on empirical data behavior descriptions [Li *et al.*, 1998; Li and Smith, 2000]. An example of the use of these system components is model-based mining of diverse data sources for predicting locations of disease outbreak and spread of insect infestation. In this example, a set of models based on mining the geo-spatial (e.g. population density, climactic, environmental) and temporal characteristics of the epidemiological knowledge of the infectious disease was developed [Glass *et al.*, 2000].

Our proposed system would include adaptation of XML Integrator (XI) and Application Roaming Framework components to handle query specialization and schema for data mapping. The XML integrator provides a unified framework to address both syntactic and semantic interoperability issues originated from the necessity of having to federate heterogeneous collections of data sources. The syntactic federation is addressed by capturing the unified data model, while the semantic federation is addressed by providing infrastructure to capture the ontology and taxonomy arising in the data sources. The Application Roaming Framework sequences the distributed data processing flow among different repositories. An additional component will enable the optimal planning of the data processing, mining, and query among distributed data sources (OLAMP, On-Line Analysis and Mining Processing).

The data layer of a grid system would have several advantages over the current more traditional physical data archive system, e.g. in maintaining author control of data sources, improving timeliness in data availability (over the typical delays in physical transfer of data to an archive location), allowing multiple data models/formats (rather than requiring standardization), insulating data users from the technical aspects of data use, minimizing duplication of data storage resources, and reducing data manager time to physically integrate data sources. The information layer will incorporate an ontology for accessing diverse information relevant to substance abuse-related research (see additional discussion below).

Model construction will be an essential component of the proposed system (within the OLAMP component) [Hobbs, 2001; Hosking *et al.*, 1997; Lin *et al.*, 2001]. A very general conceptual model of drug use and treatment history will guide the development, based on treatment outcomes literature. We can consider that drug use is an episodic behavior over time impacted by and/or related to many domains of personal and contextual characteristics, including treatment (which is itself episodic over time), and other behaviors (such as criminal behaviors). Some domains may be represented at the individual user level; others represent characteristics at a more aggregate treatment program or system level. The model development process will be interactive (researcher/theory/data) in order to identify domains in the general model that are represented in the available data sources, indicators of those domains, functional forms of relationships (e.g. linear/non-linear), and dynamic relationships to produce analytic models that are adequate, according to model test and comparison criteria.

As part of development, we must address field site (data source and user sites) technical requirements for implementation. Issues include: adaptation of software interfaces for specific field computer systems, de-identification requirements, and user training. System performance must be assessed using the metrics including scalability, usability, sustainability, accuracy, and utility of the potential user input will be solicited to help guide the development and to assess system usability and utility.

4. Challenges in System Development and Implementation

Several areas present challenges in our development and implementation process. Interestingly, most of the technical aspects of the proposed development and implementation are not expected to present barriers since components have been successfully adapted for other health-related applications. Challenges include ontology development, several issues related to the context and culture within which the applications must work, and cost-effectiveness; similar challenges have been identified in other fields, e.g. bioinformatics, but their resolution must consider the substance abuse context (e.g. Chung and Wooley, 2003).

4.1. *Ontology development*

Ontologies must be created to represent critical client, treatment, contextual, and other domains important in assessing treatment and prevention services context, needs, and outcomes and to reflect data structures. Generalized schema do not currently exist at the level of detail needed for substance abuse-related data. However, several initiatives form a basis for domain structure development. For example, NOMS defines basic dimensions and data elements to be included in collection and reporting of data through funding and oversight by the (federal) Substance Abuse and Mental Health Administration (2004b). The NIDA Clinical Trials Network adopted a core set of data elements with standard definitions. Possible domains can be identified from the treatment outcomes literature. In addition, an example of a beginning framework was developed by UCLA Integrated Substance Abuse Programs (ISAP) to represent their current and recent longitudinal studies. A challenge is to quickly develop a usable ontological structure of drug use and treatment processes that is applicable in the short term at least for a subset of available data sources, is flexible and extensible to additional data sources, and is integrable with ontologies from complementary disciplines such as medicine and/or biology. Software, guidelines, and examples exist to facilitate ontology building [Gennari *et al.*, 2002; Martone *et al.*, 2004; Sim *et al.*, 2004; Stevens *et al.*, 2002; Wroe *et al.*, 2003].

4.2. Data source control, proprietary issues

With data from heterogeneous and autonomous agencies and entities, there exists an issue of control of data and of the results of data mining. In some states, health and social service agencies function under a single administrative structure or are cooperatively linked with shared data access, thus minimizing the number of entities involved in data access negotiations. But in other states, each agency is a separate bureaucracy and data access must be negotiated with each separately. Such negotiation processes are not insurmountable but can require considerable time, effort, and patience (e.g. Hser *et al.*, 2003). The requirement by some federal agencies of data sharing as a requirement of research project funding may facilitate access to some data sources (National Institutes of Health [NIH], 2004).

4.3. Privacy, security issues

Data privacy issues as required by HIPAA and other regulations are a major concern of potential data contributors [Berman, 2002; NIH, 2004]. Security technology and software exist and are being adapted to grid infrastructures which must conform to HIPAA and Institutional Review Board policies (e.g. Ellisman and Reilley, 2004), but local concerns currently remain about effectiveness of such technology. The proposed system must be designed to access data elements in on-going administrative databases, with primary control of data privacy remaining with the originating source agency. Accordingly, levels of user access and levels of data elements accessible must be agreed on and then implemented and monitored within the system. For example, person identifiers would not be retrievable by system users. The system's insulation of its users from actual databases (with layers of system structure and security between) is one of the advantages of our proposed system.

4.4. Costs to implement/maintain system

Most substance abuse data and potential users come from agencies/entities with limited resources and conflicting priorities for those resources. Technical expertise and computer resources vary considerably. Potential investment in major system change typically requires considerable justification of cost effectiveness. Costs have been identified as a barrier to implementing drug treatment and outcomes monitoring systems (Brown *et al.*, 2003; Hser *et al.*, in press). We anticipate the need to stage the development efforts, first implementing a prototype system to illustrate applicability and allow estimation of costs of implementation and maintenance.

4.5. Historical hypothesis-testing paradigm

The primary paradigm for analysis in substance abuse research has historically been that of hypothesis-testing. This has led to some constraints in the readiness to embrace broader data mining approaches. While data mining has been used in related areas of epidemic surveillance, genetic research, neuro-imaging, and geo-mapping for crime interdiction (General Accounting Office, 2004), its application is limited in much of substance abuse research and evaluation. We foresee the broader data mining approach (as opposed to traditional hypothesis-testing) as allowing additional exploration and model building as a basis for generating hypotheses. We hope that implementation of a prototype system will provide concrete examples to illustrate to potential users the utility of the approach.

4.6. Utility, usability, credibility of such a system

Because of the various considerations/challenges mentioned, many potential contributors and users of a data access and mining system for substance abuse research require credible evidence of the utility and usability of such a system (Brown *et al.*, 2003; Hser *et al.*, 2003). We expect to be able to provide such evidence from a staged development and implementation process. We must document system utility along several metrics, including: efficiency, represented by retrieval speed and volume processed; scalability, represented by number of users and amount of data that the system can handle; usability, represented by user satisfaction, actual use, and user time required for system use; accuracy of results (precision), representing the congruence of traditional analysis approaches and system-generated results; sustainability, represented by estimated costs; utility of model results for practical application in policy, clinical practice, or research. An additional system metric is, not just its capabilities, but its actual use to support policy and clinical decisions. A longer term measure of system success would be substantive impact on the drug abuse problem, e.g. reduction in number of users, decrease in interdiction costs, improvement in treatment outcomes.

To enhance our capability of producing a useful and credible system, we anticipate creating an advisory board composed of technical and varied potential users to provide input and monitoring during the development process.

4.7. *Funding of system development*

Another on-going challenge is securing funding for development efforts. As mentioned above, there are competing priorities for scarce resources. We are exploring possible funding mechanisms. We hope that the potential benefits of improved data access and data mining will provide justification for investment in the development process.

5. Summary

We are enthusiastic about the potential impact of expanded data access and data mining applications in the field of drug abuse research, allowing the field to explore additional complexities within a dynamic multi-faceted phenomenon. Technology is already available for such a multi-modal system and has been adapted to and is in use in other disciplines. But challenges remain in its application to drug abuse research and evaluation, including: ontology development; data control, privacy, and security; costs of development, use, and system maintenance; analysis paradigm expansion; and documenting utility, usability, and credibility to potential users. Our expected approach to meeting these challenges includes the staging of development such that a prototype small-scale system can illustrate feasibility and credibility and will serve as a flexible basis allowing expansion as more data sources and users become available.

References

Anglin, M. D., Caulkins, J., and Hser, Y.-H. (1993). Prevalence estimation: policy needs, current status, and future potential. *Journal of Drug Issues*, 23, pp. 345-360.

Berman, J. (2002). Confidentiality issues for medical data miners, *Artificial Intelligence in Medicine*, 26, pp. 25-36.

Brown, T., Topp, J., and Ross, D. (2003). Rationales, obstacles and strategies for local outcome monitoring systems in substance abuse treatment settings. *Journal of Substance Abuse Treatment*, 24, 31-42.

Camp, J., Krakow, M., McCarty, D., and Argeriou, M. (1992). Substance abuse treatment management information systems: balancing federal, state, and service provider needs. *Journal of Mental Health Administration*, 19, pp. 5-20.

Center for Substance Abuse Prevention (CSAP). (2002). *Bringing Effective Prevention to Every Community: Update on Progress in Identifying CSAP Model Programs, Synthesizing Research Findings, and Disseminating Knowledge*. Rockville, MD: U.S. Departemt of Health and Human Services, Substance Abuse and mental Health Services Administration.

Chung, S. and Wooley, J. (2003). Challenges faced in the integration of biological information. In Lacroix and Critchlow (ed.) *Bioinformatics: Managing Scientific Data*. San Francisco, CA: Morgan Kaufmann.

Coalition for Evidence-Based Policy. (2003). *Bringing Evidence-Driven Progress to Crime and Substance-Abuse Policy: A Recommended Federal Strategy. Washington, DC: The Council for Excellence in Government*. Available at http://www.excelgov.org/displayContent.asp?Keyword=prppcPrevent (Accessed 8/17/2005)

Daleiden, E. and Chorpita, B. (2005). From data to wisdom: quality improvement strategies supporting large-scale implementation of evidenced-based services. *Child and Adolescent Psychiatric Clinics of North America*, 14, pp. 329-349.

Ellisman, M. and Reilley, C. (2004). BIRN grid collaborations will enhance security, performance, and scalability. *BIRNing Issues*. 3, pp. 2-3. Available at http://www.nbirn.net (Accessed 8/17/2005)

General Accounting Office (2004). *Data Mining: Federal Efforts Cover a Wide Range of Uses. Report GAO-04-548.* Washington, DC: General Accounting Office. http://www.gao.gov/new.items/d04548.pdf (Accessed 8/17/2005)

Gennari, J., Musen, M., Fergerson, R., Grosso, W., Crubézy, M., Eriksson, H., Noy, N. and Tu, S. (2002). *The Evolution of Protégé: An Environment for Knowledge-Based Systems Development.* Stanford, CA: Stanford University.

Grella, C., Hser, Y.-I., Teruya, C., & Evans, E. (2005). How can research-based findings be used to improve practice? Perspectives from participants in a statewide outcomes monitoring study. *Journal of Drug Issues,* Summer, 465-480.

Glass, G., Cheek, J. Patz, J., Shields, T., Doyle, T., Thoroughman, D., Hunt, D., Enscore, R., Gage, K., Irland, C., Peters, C. and Bryan, R. (2000). Using remotely sensed data to identify areas at risk for hantavirus pulmonary syndrome. *Emerging Infectious Diseases,* 6, pp. 238-247.

Grethe, J., Baru, C., Gupta, A., James, M., Ludaescher, B., Martone, M., Papadopoulos, P., Peltier, S., Rajasekar, A., Santini, S., Zaslavsky, I., & Ellisman, M. (2005). Biomedical informatics research network: building a national collaboratory to hasten the derivation of new understanding and treatment of disease. *Studies in Health Technology and Informatics,* 112, pp. 100-109.

Hernandez, V. and Blanquer, I. (2005). The Grid as a healthcare provision tool. *Methods of Information in Medicine,* 44, pp. 144-148.

Hobbs, F. (2001). Data mining and healthcare informatics. *American Journal of Health Behavior,* 25, pp. 285-289.

Hosking, J., Pednault, E. and Sudan, M. (1997). A statistical perspective on data mining, *Future Generation Computer Systems,* 13, pp. 117-134.

Hser, Y.-I. (1993). Data sources: problems and issues. *Journal of Drug Issues,* 23, pp. 217-228.

Hser, Y.-I., Evans, E., Teruya, C., Ettner, S., Hardy, M. *et al.* (2003). *The California Treatment Outcome Project (CalTOP) Final Report.* Los Angeles CA: UCLA Integrated Substance Abuse Programs.

Hser, H.-I., Grella, C., Teruya, C., & Evans, E. (in press). Utilization and outcomes of mental health services among patients in drug treatment. *Journal of Addictive Diseases.*

Jindal, B. and Winstead, D. (2002). *Status Report on Research on the Outcomes of Welfare Reform.* Washington, DC: U.S. Department of Health and Human Services, Office of the Assistant Secretary for Planning and Evaluation. Available at http://aspe.hhs.gov/hsp/welf-ref-outcomes02/index.htm (Accessed 8/17/2005)

Lacroix, Z. and Critchlow, T. (2003). *Bioinformatics: Managing Scientific Data.* San Francisco, CA: Morgan Kaufmann.

Lamb, S., Greenlick, M. and McCarty, D. (Eds.) (1998). *Bridging the Gap Between Practice and Research.* Washington, DC: National Academy Press.

Leshner, A. (1999). Science-based views of drug addiction and its treatment. *Journal of the American Medical Association.* 282, pp. 1314-1316

Li, C. S., Bergman, L., Castelli, V., Smith, J., Thomasian, A., Lele, S., Patz, A. and Glass, G. (1998). Model-based mining of remotely sensed data for environmental and public health applications. In S. Wong (Ed.), *Medical Image Database.* Dordrecht, The Netherlands: Kluwer Academic.

Li, C. S. and Smith, J. (2000). *An Inferencing Framework for Multi-modal Content-based Queries.* Invited paper for Conferences on Information Science and Systems, Princeton, March.

Lin, F., Chou, S., Pan, S. and Chen, Y. (2001). Mining time dependency patterns in clinical pathways, *International Journal of Medical Informatics,* 62, pp. 11-25.

Longshore, D., Urada, D., Evans, E., Hser, Y.-I., Prendergast, M. and Hawkins, A. (2005). *Evaluation of the Substance Abuse and Crime Prevention Act, 2004 Report.* Los Angeles CA: UCLA Integrated Substance Abuse Programs.

Martone, M., Gupta, A., and Ellisman, M. (2004) e-Neuroscience: challenges and triumphs in integrating distributed data from molecules to brains. *Nature Neuroscience,* 7(5), pp. 467-472.

National Institutes of Health (NIH). (2004). *Data Sharing Workbook.* Available at http://grants.nih.gov/grants/policy/data_sharing/data_sharing_workbook.pdf (Accessed 8/18/2005)

Office of National Drug Control Policy (2004). The Economic Costs of Drug Abuse in the United States 1992-2002. Washington, DC: Executive Office of the President (Publication No. 207303). http://www.whitehousedrugpolicy.gov/publications/pdf/ economic_costs98.pdf (Accessed 8/17/2005)

Seifert, J. (2005). Data Mining: An Overview. Congressional Research Service Report for Congress RL31798. Washington, DC: Library of Congress. http://www.fas.org/irp/crs/RL31798.pdf (Accessed 8/18/2005)

Sim, I., Olasov, B., and Carini, S. (2004). An ontology of randomized controlled trials for evidence-based practice: content specification and evaluation using the competency decomposition method, *Journal of Biomedical Informatics,* 37, pp. 108-119.

Simpson, D. (2002). A conceptual framework for transferring research to practice. *Journal of Substance Abuse Treatment,* 22, 171-182.

Stevens, R., Goble, C., Horrocks, I. and Bechhofer, S. (2002). Building a bioinformatics ontology using OIL, *IEEE Transactions in Information Technology in Biomedicine,* 6, pp. 135-141.

Substance Abuse and Mental Health Services Administration. (2004a). *Results from the 2003 National Survey on Drug Use and Health: National Findings* (Office of Applied Studies, NSDUH Series H–25, DHHS Publication No. SMA 04–3964). Rockville, MD.

Substance Abuse and Mental Health Administration (2004b). *Strategic Prevention Framework.* Rockville, MD: U.S. Department of Health and Human Services. Available at http://www.samhsa.gov/Matrix/SAP_prevention.aspx (Accessed 8/19/ 2005)

Substance Abuse and Mental Health Administration (2005). Measuring Outcomes to Improve Services. SAMHSA News, 13(4). Available at http://alt.samhsa.gov/samhsa_news/VolumeXIII_4/article9.htm (Accessed 8/20/2005) (See also http://www.nationaloutcomemeasures.samhsa.gov/)

Wroe, C., Stevens, R., Goble, C., Roberts, A., and Greenwood, M. (2003). A suite of DAML+OIL ontologies to describe bioinformatics web services and data. *International Journal of Cooperative Information Systems*, 12, pp. 197-224.

CHAPTER 10

KNOWLEDGE REPRESENTATION FOR VERSATILE HYBRID INTELLIGENT PROCESSING APPLIED IN PREDICTIVE TOXICOLOGY

Daniel NEAGU[1], Marian CRACIUN[2], Qasim CHAUDHRY[3], Nick PRICE[4]

[1]*Department of Computing, School of Informatics, University of Bradford, BD7 1DP, United Kingdom*

[2]*Department of Computer Science and Engineering, University "Dunarea de Jos" of Galati, Romania*

[3]*Central Science Laboratory, Sand Hutton, York YO41 1 LZ, United Kingdom*

[4]*Technology for Growth, Oak Lea, Marton, Sinnington, North Yorkshire Y062 6RD, United Kingdom*

Abstract

The increasing amount and complexity of data used in predictive toxicology call for new and flexible approaches based on hybrid intelligent methods to mine the data. This chapter introduces the specification language PToxML, based on the open XML standard, for toxicology data structures, and the markup language HISML for integrated data structures of Hybrid Intelligent Systems. The second XML application was introduced to fill the gap between existing Predictive Toxicology simple models and complex models based on explicit and implicit knowledge represented as modular hybrid intelligent structures. The two proposed specification languages are of immediate use for predictive toxicology and soft computing applications. First results to represent and process specific features of two pesticides groups through clusters and modular hybrid intelligent models using PToxML and HISML are also presented.

1. Introduction

The main aim of this chapter is to explore knowledge representation issues regarding integration of Artificial Intelligence (AI) structures to propose and validate robust models to predict, mine and exploit toxicological data.

The research effort to model and process predictive toxicology knowledge requires more flexible standards because of the continuously increasing number of descriptors, endpoints and classifiers. Any modeling approaches in Predictive Toxicology (PT) require a stepwise integration of analytical stages between data collection and models development. Computational PT approaches follow the steps: data pre-processing and filtering; feature evaluation and selection; model development and evaluation; knowledge extraction; adaptation to new data. Many of such simple components are regularly reviewed and updated. Consequently, for easy access, homogeneous updates and data dissemination, and also for integration of soft computing models with other information processing domains (chemoinformatics, bioinformatics, bio-chemistry), suitable data structures are needed.

Our models, based on an original approach using Hybrid Intelligent Systems (HIS), Modular Neural Networks (MNN) and Multi-Classifier Systems (MCS), are based on an integrated AI platform to combine experimental validation with *in silico* toxicity profiling of chemical compounds (see Section 5 of this chapter). Therefore, experts can add results from experiments and also their expertise, and general users can retrieve data and knowledge that will provide clues to improve existing predictive models. To achieve this goal, experimental data and human expertise are integrated in ensembles of connectionist and fuzzy inference systems, thereby selecting useful molecular descriptors and multiple predictive models to produce a single composite model that generally exhibits greater predictive power and generality.

Predictive Data Mining (PDM) proposes search algorithms for strong patterns in data, which can generalize accurate further decisions (Weiss and Indurkhya, 1998). The goal of PDM is to find a description of patterns and relationships in data for further predictions. In fact, PDM can be considered a Pattern Recognition task – the act of considering in raw data and taking an action based on the "category" of the pattern (Duda *et al.*, 2000). There are many Statistics and Machine Learning (ML) algorithms used in PDM: Linear Regression, Principal Component Analysis, Logistic Regression, Discriminant Analysis, k-Nearest Neighbors, Rule Induction, Decision Trees, Neural Networks, Fuzzy Inference Systems etc.

Predictive Toxicology (PT) is a multi-disciplinary science that requires a close collaboration between researchers from Toxicology, Chemistry, Biology, Statistics and Artificial Intelligence (Woo, 2001). The domain of our study is the prediction of toxicity for molecules, an investigation area rapidly growing from chemometrics and data mining. PT aims to describe relationships between chemical structures and biological/ toxicological processes: Structure–Activity Relationships (SAR). Since there is no *a priori* information about existing relationships between chemical structures and biological effects, development of a predictive model is based on patterns determined from available data. This is the framework that brings together PDM and PT.

Many factors and parameters are considered in the toxicity mechanism. No single property can satisfy the requirement to model alone toxicity, though one can identify some interesting cases, such as logP to describe narcosis (Bradbury and Lipnick, 1990). Thus, large number of features can normally be tested. There is also a representation problem. Many molecular representations have been proposed, claiming to explain properties of the molecule better (quantum similarity, spectral properties, descriptors etc.), but no general conclusions are reached. The main objective of PT is to work "*in silico*, not *in vivo*": to assess toxicity in a virtual laboratory and to propose as general models as possible for large groups of compounds and endpoints. This trend is not new in chemistry: computational models are present in any area of chemical analysis and synthesis.

Therefore, PDM in Toxicology requires developing an integrated knowledge representation approach:

- to assure a translational bridge for available data and knowledge between various software tools and file formats one can use nowadays;
- to propose a valid model to collect, update, interrogate, and authoring knowledge using common ways for broad access (including distributed web databases);
- to propose an unified framework to integrate various AI tools to process toxicological, biological and chemical data using flexible modular models in order to represent distributed quantitative implicit knowledge from experimental databases and qualitative explicit knowledge of human experts (Neagu and Gini, 2003).

There are some general data structures currently available for data storage for some particular machine learning models, and also some attempts to represent chemical compounds information as XML files. Thus, Section 2 briefly presents Machine Learning Techniques and Hybrid Intelligent Systems currently used in knowledge representation and processing for Predictive Toxicology data. Section 3 reviews existing XML specifications for standard data representation in AI, Chemoinformatics and Predictive Toxicology. For toxicity prediction, a way to manage noisy, inconsistent or missing data from various sources is to employ an ontological approach and group the compounds into homogeneous sets. Section 4 presents the proposed specification language PToxML for toxicology data structures based on the open XML standard. In Section 5, HISML - a XML scheme for Hybrid Intelligent Systems is proposed. First results to represent and process specific features of two pesticides groups using PToxML and HISML are reviewed in Section 6. The chapter ends with some conclusions and directions for further research.

2. Hybrid Intelligent Techniques for Predictive Toxicology Knowledge Representation

In the area of toxicity prediction and Quantitative SAR (QSAR) modeling, various AI techniques were proposed: Decision and Regression Trees, Classification Rules, Inductive Logic Programming (ILP), Artificial Neural Networks (ANN) or hybrid approaches like Neuro-Fuzzy models. There were also approaches based on competitive or cooperative combinations of two or more global or local techniques rather than using a single approach. In Section 5 of this chapter will be presented details of an original hybrid intelligent system for toxicity prediction, which integrates the knowledge represented by two main concepts (Neagu and Palade, 2002): implicit knowledge and explicit knowledge.

Many types of ANNs are used to represent PT knowledge and to model data: supervised ANNs, such as Backpropagation and Radial Basis Function Networks, Generalized Regression Networks, Probabilistic Neural Networks, and also unsupervised ANNs, such as Self-Organizing Maps and Learning Vector Quantization Nets.

Probabilistic Neural Networks were used by Kaiser and Niculescu (2001) as the basis for two stand-alone toxicity computation programs of TerraBase Inc. to compute the acute toxicity of chemicals to Daphnia Magna. Backpropagation ANNs are used by (Gini and Lorenzini, 1999) to predict carcinogenicity of aromatic nitrogen compounds, based on molecular descriptors: electrostatic, topological, quantum-chemical, physicochemical, etc. ANNs were also used by (Bahler *et al.*, 2000) for multiple tasks: to build a tool for predicting chemically induced carcinogenesis in rodents and to find a relevant subset of attributes (feature extraction) using a so called Single Hidden Unit Method.

In Structure-Activity Relationship problems, Artificial Neural Networks were studied also by (Adamczak and Duch, 2000): logical rules are extracted in order to describe data from two SAR sets: antibiotic activity of pyrimidine compounds, and carcinogenicity data from the Predictive-Toxicology Evaluation project of the US National Institute of Environmental Health Science (NIEHS).

Learning Vector Quantization (LVQ) based models for Predictive Toxicology are presented in (Baurin *et al.*, 2001): 4 LVQ sub-optimal models on 4 sex/species combinations (male mouse, female mouse, male rat, female rat) for carcinogenicity prediction, with fingerprints (i.e. the presence or absence within a molecule of various structural features) as inputs for each molecule.

In (Pintore *et al.*, 2001), after the reduction of the number of input variables using Genetic Algorithms, a hybrid system combining SOM and Fuzzy Clustering is applied for central nervous system (CNS) active compounds. A Fuzzy Logic procedure called Adaptive Fuzzy Partition (AFP) – a supervised classification method based on a fuzzy partition algorithm, is applied by Pintore *et al.* (2002) to a pesticides data set. A neuro-fuzzy system called Feature Space Mapping was used in (Adamczak and Duch, 2000): the results were compared with other methods (a constrained Multi-Layer Perceptron used to extract logical rules, Golem – an ILP system, standard linear regression and CART — a decision tree algorithm).

3. XML Schemas for Knowledge Representation and Processing in AI and Predictive Toxicology

Solving problems using AI techniques requires various specialized languages, such as Lisp and Prolog for Logic Programming, or programming languages, such as Matlab®, Java™, C/C++, to implement Machine Learning algorithms. Moreover, another issue in AI applications started to claim more attention nowadays: data and knowledge representation. A unified way to propose a framework for general standard specification of data contents and structure is XML. This opens the possibilities to develop domain dictionaries, robust data processing and validation by metadata definition. There are some attempts so far to encapsulate knowledge, rules or data which could be further used to create and represent AI systems.

Tabet *et al.* (2000) show that the input and output, and even the rules themselves, from an AI application can be given as XML files: this allows reducing considerable time and effort in building conversion procedures. They work to develop Universal Rule Markup Language

(URML) with the goal to promote the development of standards for rule markup using XML.

An original communication language and protocol for knowledge exchange for intelligent information agents, Knowledge Query and Manipulation Language (KQML) offers a useful abstract level for definition of distributed AI systems. KQML can be used as a language for an application program to interact with an intelligent system, as well as for two or more intelligent systems to share knowledge for cooperative problem solving (Finin *et al.*, 1997).

Formal Language for Business Communication (FLBC), a competitor to KQML, is a XML-based formal language proposed by Scott A. Moore, which one can use it for automated electronic communication.

The DARPA Agent Markup Language (DAML), developed as an extension to XML and Resource Description Framework (RDF), provides a basic infrastructure that allows a machine to make the same sorts of simple inferences that human beings do.

Case Based Markup Language (CBML) is an XML application for data represented as cases to facilitate knowledge and data markup readily reusable by intelligent agents (Hayes and Cunningham, 1998). Another effort in this direction is Artificial Intelligence Markup Language (AIML), an XML-based language used in ALICE, a chat-bot. This markup language offers a simple yet specialized open-source representation alternative for conversational agents.

For specialized Machine Learning techniques, the Predictive Model Markup Language (PMML) is a language proposed to describe Statistical and Machine Learning — Data Mining models. PMML describes the inputs to data mining models, the transformations used prior to prepare data for data mining, and the parameters which define the models themselves.

On the other side, Predictive Toxicology requires, besides contributions from Statistics and Artificial Intelligence (Data Mining, Machine Learning), a close collaboration with researchers from Chemistry, Biology, Toxicology (Woo, 2001). In all these domains the standardization of data types is an important issue and there are already some existing approaches in real world applications.

An open, platform-independent, uniform and available free of charge set of standards are proposed by Chem eStandards™, developed specifically for buying, selling and delivery of chemicals.

A new approach to manage molecular information, covering macromolecular sequences to inorganic molecules and quantum chemistry is given by Chemical Markup Language (CML). It is successfully used in conversion of chemical files without semantic loss into structured documents, including chemical publications, and a precise location of information within files.

The scientists at Leadscope® Inc. are currently leading a public initiative to promote the efforts to develop the "first toxicology controlled vocabulary for data mining": ToxML is intended to evaluate and integrate data from diverse sources in order to enable the possibility of assessment of toxicity by statistical or artificial intelligence computational models.

4. Towards a Standard for Chemical Data Representation in Predictive Toxicology

Chemical compounds can be identified in a number of different ways. For example, common practice for regulatory bodies and chemists is to use a so-called Chemical Abstract Service (CAS) number, a chemical formula, one or more chemical name(s) and a list of different chemical descriptors (e.g. structure, physico-chemical properties etc.). On the other hand, toxicologists associate a chemical compound with its effect on the life forms (plants, animals, human beings), being interested in negative effects the chemical can have. Experimental data usually represent chemicals associated with a certain mechanism of action and a certain dose for toxic effects in certain conditions on some target species.

To gather and evaluate the information for development of computational models capable to assess the toxic effect of chemicals using structural descriptors is a difficult task, given distributed and heterogeneous data sources. An important step then is the initiative of PT data and vocabulary standardization, where XML provides a powerful way to describe objects (chemicals in our case).

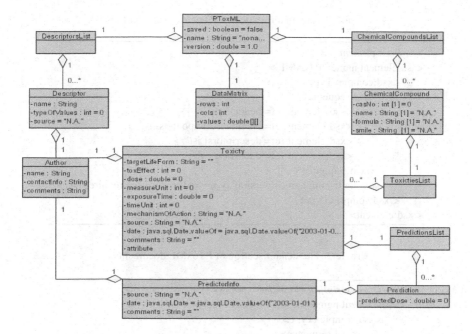

Figure 1: The architecture of PToxML.

As descriptors generating software platforms and modeling techniques vary considerably, XML promises to be the most likely language for representation of toxicology data and predictive models. We propose PToxML (Fig. 1) as an XML application to describe chemical information related to predictive toxicology.

The simplest way to describe predictive toxicology data is to present the list of chemical compounds, the list of calculated descriptors, and the matrix of numerical, categorical or missing values corresponding to each descriptor for each chemical compound. Thus, PToxML documents consist of three main sections: the header section, the chemical identification section and the data section (Table 1).

The header section defines authoring information (version of the dataset, calendar date, author/owner, initial source(s) and comments, see Table 2), whereas the second section relates properties to entities (chemical compounds list and descriptors list) and the third section presents data rows.

Table 1. The main section of PToxML documents.

```
<!—Root element -->
<xsd:element name='PToxML'>
    <xsd:complexType>
        <xsd:sequence>
            <xsd:element ref='header' />
            <xsd:element ref='chemicalCompoundsList' />
            <xsd:element ref='descriptorsList' />
            <xsd:element ref='dataMatrix' />
        </xsd:sequence>
        <xsd:attribute name='version' type='xsd:string' use='required'/>
    </xsd:complexType>
</xsd:element>
```

Table 2. The header section of PToxML documents.

```
<!—The description of header -->
<xsd:element name='header' id='el.header'>
    <xsd:complexType>
        <xsd:sequence>
            <xsd:element ref='version' />
            <xsd:element ref='date' />
            <xsd:element ref='author' />
            <xsd:element ref='source' />
            <xsd:element ref='comments' minOccurs='0'/>
        </xsd:sequence>
    </xsd:complexType>
</xsd:element>
```

The chemical compounds list is a simple unordered list of zero or more chemical compounds characterized by the required attribute 'length'. The concordance between this attribute and the number of chemicals presented in the list must be verified by the application using this data. Practically, chemical identification section (compounds list and descriptors list) provides the meta-structure of quantitative data (compounds as rows and descriptors as columns): the matrix of values in PToxML data section. Such organization of PToxML documents permits further correlation checks, i.e. the chemical compounds number is equal with the matrix rows number, whereas the number of descriptors gives the columns size.

Table 3. The structure of a chemical compound.

```
<!-- The description of a chemical compound -->
<xsd:element name='chemicalCompound' id='el.chemicalCompound'>
    <xsd:complexType>
        <xsd:sequence>
            <xsd:element ref='CASNo' />
            <xsd:element ref='Name' />
            <xsd:element ref='Formula' minOccurs='0' />
            <xsd:element ref='SMILE' minOccurs='0' />
            <xsd:element ref='toxicity' minOccurs='0'/>
        </xsd:sequence>
    </xsd:complexType>
</xsd:element>
```

A chemical compound is identified by a 'CASNo' and a 'Name' (both required); there are also chemical 'Formula', 'SMILES' code and 'toxicity' (all optional, Table 3). SMILES stand for Simplified Molecular Input Line Entry Specification, widely used as a general-purpose chemical nomenclature and data exchange format.

The toxicity structure is one of the main features of a PToxML document (Table 4). The information on toxicity (possibly unavailable for some chemicals) contains a list with one ore more 'toxInfo' elements to describe: species (target life form) part of the experiments, measured toxic effects (e.g. LC50 - lethal concentration for 50% of a population) - a compulsory field, mechanism of action (if available), toxic dose (if available), measurement unit (mg/l, mmol/l etc.), exposure time for the life form and time units. Information about source and date when this data was published, together with comments and name of the authors, are also supplied. Prediction information is a list of "prediction" fields: predicted values and predictor information (the model used to obtain the value, the source of the model, date, comments and author).

Finally, the chemical descriptors are defined by their name, value type (numerical/categorical), source and some authoring data (Table 5).

Table 4. The toxicity and prediction structures.

```
<!-- The toxicity related information -->
<xsd:element name='toxInfo' id='el.toxInfo'>
    <xsd:complexType>
        <xsd:sequence>
            <xsd:element ref='targetLifeForm' />
            <xsd:element ref='toxEffect' />
            <xsd:element ref='dose' minOccurs='0' />
            <xsd:element ref='measureUnit' />
            <xsd:element ref='exposureTime' />
            <xsd:element ref='timeUnits' />
            <xsd:element ref='mechanismOfAction' minOccurs='0' />
            <xsd:element ref='source' />
            <xsd:element ref='date' />
            <xsd:element ref='comments' minOccurs='0' />
            <xsd:element ref='author' minOccurs='0' />
            <xsd:element ref='prediction' minOccurs='0'
                            maxOccurs='unbounded'/>
        </xsd:sequence>
    </xsd:complexType>
</xsd:element>
<!-- The prediction  related information -->
<xsd:element name='prediction' id='el.prediction'>
    <xsd:complexType>
        <xsd:sequence>
            <xsd:element ref='predictedDose' />
            <xsd:element ref='predictorInfo' />
        </xsd:sequence>
    </xsd:complexType>
</xsd:element>
```

Table 5. The structure of a descriptor.

```
<!--Descriptors related information -->
  <xsd:element name='descriptor'>
    <xsd:complexType>
      <xsd:sequence>
        <xsd:element ref='name' />
        <xsd:element ref='typeOfValues' />
        <xsd:element ref='source' />
        <xsd:element ref='author' />
      </xsd:sequence>
    </xsd:complexType>
  </xsd:element>
```

5. Hybrid Intelligent Systems for Knowledge Representation in Predictive Toxicology

The last ten years have produced a tremendous amount of research on fuzzy logic and connectionist systems. The two approaches can be used in a complementary way, HIS combining their features (Fig. 2). In such systems, the learner can insert fuzzy rules into neural networks. Once the domain knowledge has a neural representation, training examples are used to refine initial knowledge or additional structures. Finally, it processes the output for given instances and, using specific methods (Benitez *et al.*, 1997; Neagu and Palade, 2000), can extract symbolic information from trained networks, to explain and interpret the refined connectionist knowledge.

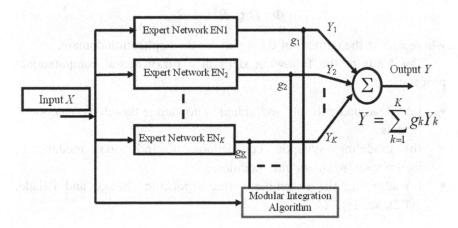

Figure 2: The architecture of an integrated implicit and explicit knowledge-based system.

The *implicit knowledge* is defined as connectionist representation of learning data. An *explicit knowledge* module has the role to adjust performances of implicit knowledge modules by using external information provided by experts, in form of Fuzzy Rule-based Systems. In our approach, connectionist integration of explicit and implicit knowledge appears a natural solution to develop homogeneous intelligent systems (Neagu and Palade, 2002). Explicit and implicit rules are represented using MLP (Multi-Layer Perceptron, Rumelhart and McClelland, 1998) Crisp Neural Networks (CNN), neuro-fuzzy or fuzzy (FNN) neural nets (Buckley and Hayashi, 1995). Thus, fuzzy logic provides the inference mechanism under cognitive uncertainty, since neural nets offer advantages of learning, adaptation, fault-tolerance, parallelism and generalization.

5.1. *A formal description of implicit and explicit knowledge-based intelligent systems*

The HIS considered in this work is a multi-input single-output hybrid system (MISO). Its general goal is to model data and expert information in order to relate the inputs with corresponding output values:

$$\Phi : D \subseteq R^n \to R \qquad (1)$$

where $n \in N$ is the number of the inputs for the application domain.

This leads to the following steps in a fuzzy neural computational process:

- (a) development of individual knowledge-based connectionist models,
- (b) modeling synaptic connections of individual models, to incorporate fuzziness into modules,
- (c) adjusting the ensemble voting algorithm (Neagu and Palade, 2002), see Fig. 2.

Let's consider a MISO HIS with n inputs. Let also consider $U = \prod_{i=1}^{n+1} D_i$ as the universe of discourse over the application domain as the Cartesian product of sets D_i, $i=1..n+1$, for the input variables $X_i \in D_i$, $i=1..n$, and the output $Y \in D_{n+1}$. A HIS integrated model of the problem Φ based on implicit (IKM) and explicit knowledge (EKM) modules, is a good approximation of Φ if:

$$HIS = \left\{ M_j \underset{j=1,..,m}{/} \forall \varepsilon > 0, \exists X \in \prod_{i=1}^{n} D_i, \forall Y = \Phi(X) : \left\| M_j(X) - Y \right\| < \varepsilon \right\} (2)$$

where the knowledge modules are functional models:

$$M_j : \prod_{i=1}^{n} D_{ij} \rightarrow D_{n+1, j} \tag{3}$$

The modules M_j are, in our approach, either implicit or explicit knowledge models:

$$M_j \subset \{M_{IKM\_CNN}, M_{IKM\_FNN}, M_{EKM\_Mamdani}, M_{EKM\_Sugeno}\}$$

For any of these M_j models, we can propose, following Eqs (1)-(3), a formal parameter-based description of HIS:

$$M_j = \langle \Theta, \Lambda, \Omega \rangle \tag{4}$$

where Θ is the set of *topological parameters* (i.e. number of layers, number of neurons on each layer, connection matrices, type and number of individual models and gating networks), Λ is the set of *learning parameters* (learning rate, momentum term, any early stopping attribute for implicit knowledge modules, but NIL for explicit knowledge modules) and Ω is the set of *description parameters* (type of fuzzy sets, parameters of membership functions associated to linguistic variables).

Three distinctive cases to develop HIS models are identified:

Case 1: $D_j = \prod_{i=1}^{n} D_{ij}$ for all $j=1..m$: a modular architecture (Neagu and Palade, 2002) of expert on the whole input domain.

Case 2: $\bigcap_{j=1}^{m} \prod_{i=1}^{n} D_{ij} = 0$ and $D_i \cap D_j = 0$, for $j,k = 1,..,m$. The HIS model is a collection of m experts on disjunctive input domains; the system is a top-down integrated decomposition model, dividing the initial problem in separate less-complex sub-problems (Craciun and Neagu, 2003).

Case 3: $\bigcap_{j=1}^{m} \prod_{i=1}^{n} D_{ij} \neq 0$: models built on overlapping sub-domains; further algorithms to refine the problem as cases 1 or 2 are required.

Starting from basic combination strategies - simple voting or averaging (Breiman, 1996b; Freund, 1995), researchers proposed to extend the concept of ensemble development from Pattern Recognition to Artificial Intelligence, including the statistical meaning of the result and the knowledge level of the proposed combination. The meta-learning concept, combining prediction of trained individual experts (level-0 models) with a gating function (level-1 models) is representative for mixture of experts (Jacob *et al.*, 1991) or stacking strategies (Breiman, 1996a; Wolpert, 1992). In this way the inputs of the models themselves and their output are used as inputs for a meta-model $\overline{M}$ – either an implicit or an explicit module, similar with those described above:

$$\overline{M} \subset \{M_{IKM\_CNN}, M_{IKM\_FNN}, M_{EKM\_Mamdani}, M_{EKM\_Sugeno}\}$$

So far, few strategies to combine IKM and EKM in a global HIS have been proposed (Neagu and Palade, 2002): Fire Each Module (FEM), Unsupervised-trained Gating Network (UGN), Supervised-trained Gating Network (SGN), majority voting etc. FEM is an adapted Fire Each Rule method (Buckley and Hayashi, 1995) for modular networks, in two versions: statistical combination of crisp outputs (FEMS) or fuzzy inference of linguistic outputs (FEMF). UGN proposes competitive aggregation of EKMs and IKMs, while SGN uses a supervised trained layer to recognize component outputs.

5.2. *An XML schema for hybrid intelligent systems*

The aim of this section is to propose the standard XML application for knowledge representation, data exchange and analysis of AI experimental data: the Hybrid Intelligent Systems markup language (HISML). From our knowledge, this is the very first attempt to propose a standard for integrated soft computing techniques, such as HIS. The proposed HISML syntax captures the structure and parameters of modeling experiments (Fig. 3). The information stored in a HISML document is further required to analyze and replicate the models.

Table 6. The main structure of HSIML.

```
<!-- Root element -->
<xsd:element name='HISML'>
    <xsd:complexType>
        <xsd:sequence>
            <xsd:element ref='header' />
            <xsd:element ref='HIS' />
        </xsd:sequence>
        <xsd:attribute name='version' type='xsd:string' use='required'/>
    </xsd:complexType>
</xsd:element>

<!-- Header Description -->
<xsd:element name='header' id='el.header'>
    <xsd:complexType>
        <xsd:sequence>
            <xsd:element ref='author' />
            <xsd:element ref='date' />
            <xsd:element ref='comments' minOccurs='0'/>
        </xsd:sequence>
    </xsd:complexType>
</xsd:element>
```

The seed of the common format to manipulate as well as to store the data is captured by the concept of the 'HISML' element. Such an element has a required attribute 'version' and contains two other sections: a header with authoring information and section 'HIS' to define recurrent intelligent component systems (Table 6).

One of the main features the HISML syntax takes into consideration is the recurrent organization of imbricate modules. The need to propose a standard format for data storage in HIS development using the HISML syntax is justified by various methods involved (see Section 5.1) and the aim of automation for many of design steps.

A 'HIS' element comes with a 'name' attribute, and might be either a 'simpleHIS' (basic atoms IKM and EKM) or a 'complexHIS' (any other HIS, including simple ones), plus additional information about its performance (Table 7).

Table 7. The structure of a HIS.

```
<!-- Hybrid Intelligent System module description-->
<xsd:element name='HIS' id='el.HIS'>
    <xsd:complexType>
        <xsd:sequence>
            <xsd:element ref='performance' minOccurs='0'/>
            <xsd:choice>
                <xsd:element ref='simpleHIS' />
                <xsd:element ref='complexHIS' />
            </xsd:choice>
        </xsd:sequence>
        <xsd:attribute name='name' type='xsd:string' use='required' />
    </xsd:complexType>
</xsd:element>

<!-- HIS performance -->
<xsd:element name='performance' id='el.performance'>
    <xsd:complexType>
        <xsd:sequence>
            <xsd:element ref='correlationCoefficient' minOccurs='0'/>
            <xsd:element ref='meanAbsoluteError' minOccurs='0'/>
            <xsd:element ref='meanSquredError' minOccurs='0'/>
            <xsd:element ref='classificationAccuracy' minOccurs='0'/>
        </xsd:sequence>
    </xsd:complexType>
</xsd:element>
```

A 'simpleHIS' element (Fig. 3) comes with a 'name' attribute, and might be either IK or EK module. A simple HIS is finally an ANN (as IKM-CNN) or a combination of ANN with FIS (as IKM-FNN or EKM).

The basic types are ANN and FIS. An 'ANN' comes with 'train' (has been the net trained?) and topological data attributes (inputs number, hidden layers list, neurons data: activation functions, weights matrix to current layer, training algorithm). A 'FIS' includes essential information to identify fuzzy inference systems: the type of the system, also the number of inputs and outputs, the number of rules, the methods for *and*, *or*, *implication*, *aggregation* and *defuzzification*, the rules represented as matrix and the list of fuzzy variable for input and output (Table 8).

Table 8. The structure for 'FIS' – Fuzzy Inference System.

```
<!-- Fuzzy Inference System Type -->
  <xsd:complexType name='FIS' id='ct.FIS'>
    <xsd:sequence>
      <xsd:element ref='dataDescription' />
      <xsd:element ref='typeOfFIS' />
      <xsd:element ref='numberOfInputs' />
      <xsd:element ref='numberOfOutputs' />
      <xsd:element ref='numberOfRules' />
      <xsd:element ref='andMethod' />
      <xsd:element ref='orMethod' />
      <xsd:element ref='impMethod' />
      <xsd:element ref='aggMethod' />
      <xsd:element ref='defuzzMethod' />
      <xsd:element ref='inputsList' />
      <xsd:element ref='outputsList' />
      <xsd:element ref='rules' />
    </xsd:sequence>
  </xsd:complexType>
```

A 'complexHIS' (Fig. 3) consists of two or more (simple or complex) HIS elements, and a gating module 'GM' or a choosing algorithm 'CM' (either a statistical or fuzzy inference approach) as integration module to combine in an intelligent way all the modules inside. A Gating Module 'GM' can be a (supervised or unsupervised trained) CNN or a statistical combining algorithm (FEMS, majority voting, max, min, average etc.). A choice module 'CM' is a (supervised or unsupervised trained) ANN of any 'complexHIS' types.

6. A Case Study

Toxic chemicals, such as pesticides, are routinely tested on animals as part of the registration process. Current UK pesticide registration process require generation of LD50 (lethal doses for 50% of initial population) data (Home Office Statistics of Scientific Procedures on Living Animals, 1999) for at least one avian species (bobwhite quail, mallard duck, Japanese quail) to determine risks posed to wildlife. This section describes experiments on proposed XML schemas for PT models of insecticides as an alternative to the use of birds in LD50 tests.

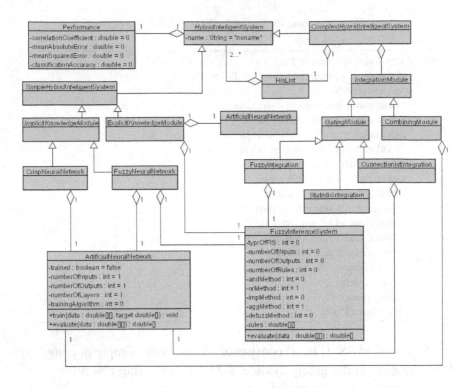

Figure 3: The architecture of HISML.

6.1. *Materials and methods*

The thoroughly quality-assured dataset of LD50 values for 84 anticholinesterase pesticides (60 organophosphorus and 24 carbamates) against mallard duck used in this study was obtained from the Canadian Wildlife Service. 3D chemical structures of the compounds were built using CHEMX (Chemical Design, UK) and CAChE (FQS Poland) modelling suites. Over 1700 molecular descriptors were generated for each optimized structure. These included 20 electronic, structural and hydrophobic molecular descriptors in the Project Leader module of CaChe 6.1, around 200 electro-topographical descriptors by RECON, and over 1500 descriptors of different sorts by DRAGON v3.0.

The entire dataset has been divided in training and independent test sets. For the test set, about 10% of compounds were extracted with respect to the distribution into the two classes. Training values were scaled to [-1,1] because of the transfer function type for the hidden layer neurons of the feed-forward back-propagation ANN (the hyperbolic tangent transfer function). The test set is scaled to the same interval using min and max values of the training set descriptors. The experiment considered a HIS implementation for Case 2 (see Section 5) of disjunctive input domains. Consequently, the data was split in two clusters before the development of a HIS model Φ:

$$\Phi(X) = \begin{cases} f_1(X), \ X \in D_1 \\ .. \\ f_K(X), \ X \in D_K \end{cases} \quad (\forall) X \in D \quad (5)$$

where $D_i \subseteq R^n$, $D = D_1 \cup .. \cup D_K$, $D_i \cap D_j = 0$, for all $i,j=1,..,K$ and $f_i=f/D_i$ are continuous projections of function f on sub-domains D_i, for $i=1,..,K$.

6.2. Results

From the original dataset, descriptors with missing values, highly cross-correlated descriptors (R^2 in excess of 0.9) and non-variant descriptors (constant values for almost chemical compounds) were discarded.

A genetic search based feature selection algorithm (Goldberg, 1989) has been applied on training set to reduce the dimensionality of the input space, using a Weka (Witten and Frank, 1999) implementation: about 88 chemical descriptors were selected for the model data set.

Once pre-processing steps done, a competitive network (self-organized neural network) as a choosing module 'CM' has been defined and learned to cluster training data in 2 disjuvnctive groups just in the input space (molecular descriptors space), based on the Euclidian distance as similarity measure. During this stage, information about output values was not used. The trained expert was able to group compounds according to their chemical classes with 85.5% accuracy on the training set and about 75% accuracy on the external test set. These results support the approach of using mathematical descriptions of chemical compounds to find the chemical class they belong to.

Table 9. Performances of disjunctive HIS models.

Dataset	R(%)	MSE	MAE
TRAIN	91.2	0.049	0.138
TEST	66.2	0.126	0.289

Table 10. Performances of disjunctive linear models.

Dataset	R(%)	MSE	MAE
TRAIN	81.1	0.363	0.465
TEST	52.6	0.198	0.354

Each cluster is further modeled by 3-layered feed-forward networks with 2 hidden units ('CNN'), linear transfer function for input and output layers, and hyperbolic tangent function for hidden layer. Each network has been trained 200 epochs using back-propagation algorithm based on gradient descent with momentum and adaptive learning rate. In Table 9 are shown values of correlation coefficient (R), mean-squared error (MSE) and mean absolute error (MAE) for training and test sets.

In order to have a reference point to report the performance of the proposed approach, two of the best linear models obtained by CSL, one for each pesticide category, are considered. Every chemical from train and test set is evaluated using the linear model corresponding to the group the unsupervised CM suggested. Because the two pesticide classes are different enough and the linear models capture well the characteristics of the category, the predicted values for the misclassified compounds are few magnitude orders far away from the values in their universe of discourse. This behavior acts like a trigger that is signaling the possibility of a wrong predicted category. In these exception cases the forecasts of the other linear model will be used instead. The values of R, MAE and MSE are shown in Table 10.

The performance of unsupervised and supervised combinations ('complexHIS') seems to be very good on the training set but poor on the test set, although special attention has being paid to avoid overfitting.

However, the prediction results are superior to those obtained by the classical linear models. An explanation could rely on the relatively small number of chemicals used in experiment and the data homogeneity, given the unequal distribution of chemicals into the classes (60 and 24, respectively).

7. Conclusions

Over the past fifteen years, schemes for structure-toxicity modeling have changed from congeneric series approach through chemical class-based approach, multiple regression and AI-based approaches. Our approach to model predictive toxicology knowledge, based on modular combination of local experts appears to propose better models. This makes our attempt challenging. One of the advantages of our approach is the flexibility of the developed models, able to be optimized for specific cases.

In this chapter, a specification language PToxML for toxicology data structures based on the open XML standard has been proposed. A XML scheme for Hybrid Intelligent Systems has been also introduced. HISML can be considered a promising standard because of the expressiveness of its models for existing soft computing techniques, and its interconnection capabilities, flexibility, and simplicity to allow general use for HIS development.

Future research steps are to test, update and (possibly) improve the overall representation of knowledge for both, PToxML and HISML. Although it might be too early to speculate on their outcomes, we propose PToxML as a gateway to exchange Predictive Toxicology data, and also to authoring and study the quality of experimental information. We also propose HISML to allow researchers a uniform access not only to soft computing techniques development, but also for further various integration and (even recurrent) combinations as HIS.

A strong point that makes combinations of classifiers attractive, beside better results in classification performance, is their ability of being distributed in time and space: inter- and intra-model parallelism (Freitas, 1998), data and processing parallelism (Taniar and Rahayu, 2002). Subsets of the data could be distributed to different processors, that apply different learning algorithms, and then results are combined to yield a single classifier. More can be achieved by harmonizing the parallel processing techniques with the combination of local and global models in ensemble experts and mixing technologies in Hybrid Intelligent Systems in order to improve the prediction accuracy and to provide reasonable training/response time.

Acknowledgements

The first author acknowledges the support of the EPSRC project PYTHIA GR/T02508/01. The authors are grateful to P. Mineau, A. Baril, B.T. Collins, J. Duffe, G. Joerman, and R. Luttik (Canadian Wildlife Service) for providing toxicity data. Acknowledgements are also due to R. Watkins (CSL) for building 3D molecular structures. The work of the second author and also the application prototype are partially funded by the EU FP5 project DEMETRA QLK5-CT-2002-00691.

References

Adamczak, R. and Duch, W. (2000). Neural Networks for Structure-Activity Relationship Problems, *Procs 5th Conf. on Neural Nets and Soft Computing*, Zakopane, pp. 669-674.

AIML: http://www.oasis-open.org/cover/aiml-ALICE.html.

Bahler, D., Stone, B., Wellington, C. and Bristol, D.W. (2000). Symbolic, Neural and Bayesian Machine Learning Models for Predicting Carcinogenicity of Chemical Compounds, *J. of Chem. Inf. and Comp. Sc.*, 8, pp. 906-914.

Baurin, N., Marot, C., Mozziconacci, J.C. and Morin-Allory L., (2001). "Using of Learning Vector Quantization and BCI fingerprints for the Predictive Toxicological Challenge 2000-2001", *Procs ECML/PKDD-01 Workshop*, PTC 2000-2001.

Benitez, J.M., Castro J.L. and Requena, I. (1997). Are ANNs Black Boxes?, *IEEE Trans Neur Nets* 8/5, pp. 1157-1164.

Bradbury, S.P. and Lipnick, R.L. (1990). Structural properties for determining mechanisms of toxic action, *Environmental Health Perspectives*, 87, pp.181-182.

Breiman, L. (1996a). Stacked generalization, *Machine Learning*, vol. 24(1), pp. 49-64.

Breiman, L. (1996b). Bagging predictors, Machine Learning, vol. 24(2), pp. 123-140.

Buckley J.J. and Hayashi, Y. (1995). Neural nets for fuzzy systems, *Fuzzy Sets and Systems* 71, pp. 265-276.

CaChe 6.1 FQS Poland: http://www.cachesoftware.com/

Chem eStandards: http://www.cidx.org/Standard/.

CML: http://xml-cml.org.

Craciun M., and Neagu, C.-D. (2003). Using Unsupervised and Supervised ANNs to study Aquatic Toxicity, *Proc. 7th Int'l Conf KES2003*, Oxford, LNAI, pp. 911-918.

DAML: http://www.daml.org/.

DEMETRA-Development of Environmental Modules for Evaluation of Toxicity of pesticide Residues in Agriculture: http://www.demetra-tox.net/.

Duda, R.O., Hart, P.E. and Stork, D.G. (2000), Pattern Classification, John Wiley and Sons, 2nd Edition.

DRAGON: (R.Todeschini, V.Consonni, A.Mauri, and M Pavan): http://www.disat.unimib.it/chm/.

Finin, T., Labrou, Y. and Mayfield, J. (1997). KQML as an agent communication language, in Jeff Bradshaw (Ed.), *Software Agents*, MIT Press, Cambridge, pp. 291-316

FLBC: http://www.oasis-open.org/cover/flbc.html.

Freitas, A.A. (1998). A Survey of Parallel Data Mining, *Proc. of the 2nd International Conference on the Practical Applications of Knowledge Discovery and Data Mining*, pp. 287-300.

Freund, Y. (1995). Boosting a weak learning algorithm by majority, *Information and Computation*, 121(2), pp. 256-285.

Gini, G. and Lorenzini, M. (1999). Predictive Carcinogenicity: A Model for Aromatic Compounds, with Nitrogen-Containing Substituents, Based on Molecular Descriptors Using an ANN, *J of Chem. Inf. and Comp. Sc.*, 39, pp. 1076-1080.

Goldberg, D.E. (1989). Genetic algorithms in search, optimization and machine learning, Addison-Wesley.

Hayes, C., and Cunningham, P. (1998). Distributed CBR using XML, *Proc. Intell Sys.&Electr. Comm Int'l Workshop*, Bremen.

Home Office Statistics of Scientific Procedures on Living Animals. (1999). Great Britain. Government Statistical Service, The Stationary Office Limited, Norwich.

Jacob, R.A., Jordan, M. I., Nowlan S. J. and Hinton, G. E. (1991). Adaptive Mixtures of Local Experts, Neural Computation, 3, pp. 79-87.

Kaiser, K.L.E. and Niculescu, S.P. (2001). Modeling the acute toxicity of chemicals to Daphnia magna: a probabilistic neural network approach, *Envir. Tox. Chem.* 20, pp. 402-431.

KQML: http://www.cs.umbc.edu/kqml/.

Neagu, D. and Palade, V. (2000). An interactive fuzzy operator used in rule extraction from neural networks, *Neural Networks World Journal* 10/4, pp. 675-684.

Neagu, D. and Palade, V. (2002). Modular neuro-fuzzy networks used in explicit and implicit knowledge integration, *Proc. 15th Int'l Conf. Florida AI Society - FLAIRS2002*, pp. 277-281.

Neagu, D. and Gini, G. (2003). Neuro-Fuzzy Knowledge Integration applied in Toxicity Prediction, in *Innovations in Knowledge Engineering* (R. Jain, A. Abraham, C. Faucher, B. Jan van der Zwaag: eds), Advanced Knowledge Int'l, pp. 311-342.

Pintore, M., Taboureau, O., Ros, F. and Chretien, J.R. (2001). Database mining applied to central nervous system (CNS) activity, *Eur. J. Med. Chem.* 36, 2001, pp. 349-359.

Pintore, M., Piclin, N., Benfenati, E., Gini, G. and Chretien, J.R. (2002). Data Mining with Adaptive Fuzzy Partition: Application to the Prediction of Pesticide Toxicity on Rats, *Env. Toxicology and Chemistry* 22/5, pp. 983-991.

PMML: http://www.dmg.org/index.html.

RECON: http://www.chem.rpi.edu/chemweb/recondoc/.

Rumelhart, D. and. McClelland, J. (1998). Parallel Distributed Processing, MIT Press.

Tabet, S., Bhogaraju, P. and Ash, D. (2000). "Using XML as a Language Interface for AI Applications", *Procs. Int'l Conf. PRICAI 2000*, LNCS 2112, Springer, pp. 103-110.

Taniar, D. and Rahayu, J. W. (2002). Parallel Data Mining, (H. A. Abbas, R. A. Sarker, C. S. Newton eds.), *Data Mining a Heuristic Approach*, Idea Group Publishing.

ToxML: http://www.leadscope.com/

URML: http://home.comcast.net/~stabet/urml.html

Weiss, S.M. and Indurkhya, N. (1998). Predictive Data Mining – a practical guide, Morgan Kaufmann Publishers.

Witten, I.H. and Frank, E. (1999). Data Mining: Practical Machine Learning Tools and Techniques with Java implementations, Morgan Kaufmann Publishers, San Francisco, CA.

Wolpert, H. (1992). Stacked Generalization, *Neural Networks*, vol 5, 1992, pp. 241-259.

Woo, Y-t. (2001). A Toxicologist's View and Evaluation, *Proc. ECML/PKDD-01 Workshop*, Predictive Toxicology Challenge 2000-2001.

CHAPTER 11

ENSEMBLE CLASSIFICATION SYSTEM IMPLEMENTATION FOR BIOMEDICAL MICROARRAY DATA

Shun Bian and Wenjia Wang

School of Computing Sciences, University of East Anglia, Norwich, NR4 7TJ, United Kingdom

Abstract

Ensemble method can be more effective when an ensemble is built by using knowledge of the diversity among base learners. However, implementation of an ensemble when evaluating diversity of learners for a given data mining task can be very time consuming. This paper presents a framework of developing a flexible software platform for building an ensemble based on the diversity measures. An ensemble classification system (ECS) has been implemented for mining biomedical data as well as general data. The ECS consists of Data Pre-process, Feature Selection, Classifiers Selection, Feature-Classifier Pair Evaluation and Selection, Combination and Decision Making. The ECS has been tested with several benchmark datasets and microarray data. The experiment results show that ECS is a practical program both in improving data mining performance and reducing computational time.

Keywords: Data mining, ensemble, classification, feature selection, microarray data.

1. Introduction

It has become clear that, in data mining, individual classifiers that work alone are less effective for real world data, but a set of classifiers that are combined in a constructive way to form an ensemble may improve their performance [Duin, 2000]. Recently, many researchers have demonstrated that using classifier ensembles indeed leads to an improved performance for many difficult problems, especially when Bayesian Network, Decision Tree and Genetic Algorithm are used as ensemble final classifier. Duin [Duin, 2000] did an experiment on combining classifiers to evaluate different combination rules, including fixed combiners and trained combiners in 2000. Cho *et al.* in 2003 [Park & Cho, 2003] proposed to make a committee with individual feature-classifier pair and use a GA-based method to search the optimal ensemble for cancer microarray dataset. Their experiment results were extremely good. Garg [Garg *et al.*, 2002] presented Bayesian Network and Tan [Tan & Gilbert, 2003] applied Bagged and Boosted Decision Tree as the final ensemble classifier. Liu [Liu & Iba, 2001] used the parallel genetic algorithm to filter out the informative genes. All these work contributed to ensemble methods for classification.

However, in practice, it still requires considerable programming work to implement ensemble systems, especially in some new applications, such as analysing microarray data. Furthermore, although some theories about ensemble system design have been explored so that the "blindness" in building system can be somehow avoided, such as finding the relationship between ensemble performance and diversity values among the base learners, experimental results are not consistent enough to suggest that these guidelines are more effective for building good ensembles.

Based on a belief that more knowledge in knowing diversity among base learners in a classification system will reveal effective ways for designing an ensemble system, this paper uses a diversity measurement to find the better learners for an ensemble classification system. An application prototype (ECS) for mining microarray data has being developed by diversity knowledge.

2. Background

2.1. *Reasons for ensemble*

There are three fundamental reasons for ensemble methods to promise reducing shortcomings of standard learning algorithms. A learning algorithm can be viewed as searching in a space of hypotheses to identify the best hypothesis in the space.

- *Statistical*: By constructing an ensemble out of all of these accurate classifiers, the algorithm can "average" their votes and reduce the risk of choosing a wrong solution.
- *Computational*: An ensemble constructed by running the local search from many different starting points may provides a better approximation to the true unknown function than any of the individual classifiers, as many learning algorithms work by performing some form of local search that may get stuck in local optima [Rivest, 1976].
- *Representational*: By forming weighted sums of hypotheses from a space of H, it may be possible to expand the space of representative functions as in most applications of machine learning the true function f cannot be represented by any of the hypotheses.

2.2. *Diversity and ensemble*

An ensemble of classifiers is a set of classifiers whose individual decisions are combined in some way (typically by weighted or unweighted voting) to produce an overall decision. A necessary and sufficient condition for an ensemble of classifiers to be more accurate than any of its individual members is that the classifiers are accurate and diverse [Hansen & Salamon, 1990].

An accurate classifier is one that has an error rate better than random guessing on new x values. Two classifiers are diverse if they make different errors on new data points [Dieterrich, 2000].

The term "diversity" in this context was firstly introduced in the software engineering [Littlewood & Miller, 1989]. The aim is to increase   the reliability of conventionally programmed solutions by combining programs that   failed   independently,   or   whose   failures were uncorrelated.   Sharkey [Sharkey & Sharkey, 1997] applied the same   idea   to   ensembles   of   neural nets so that failures on each net can be compensated by success of others. The advantage of combining nets is that "avoiding the loss of information that might result if the best performing net of a set of several were selected and the rest discarded" . By Sharkey, "a pair of nets can be said to be diverse with respect to a test set if they make different generalization errors on that test set, and not to   be   diverse   when   they   show   the   same   pattern of errors". A method to obtain the product-moment correlation coefficient between n pairs of observations, whose values are (x_i, y_i), is calculated as follows:

$$r = \frac{\sum_{i=1}^{n}(x_i - \overline{x})(y_i - \overline{y})}{\sqrt{[\sum_{i=1}^{n}(x_i - \overline{x})^2][\sum_{i=1}^{n}(y_i - \overline{y})^2]}} \quad (1.1)$$

By Eq. (1.1), when correlations are close to zero, nets are likely to share a minimum number of coincident failures, the diversity can be considered large [Sharkey & Sharkey, 1997].

Applying this idea in ensemble approach, it is intuitively accepted that the classifiers to be combined should be different, or "diverse", from each other. Therefore, diversity among the classifiers has been recognised as a key issue.

The key to the success of combining learners in machine learning is to build a set of diverse learners (or classifiers). By Tumer and Ghosh [Tumer & Ghosh, 1996], an Added Classification Error of a team under a set of assumptions is given below:

$$E_{add}^{ave} = E_{add}\left(\frac{1 + \theta(L-1)}{L}\right) \quad (1.2)$$

where, E_{add}^{ave} is the added error of the individual classifiers (all have the same error); θ is a correlation coefficient; if each input x will have L vectors of support, and the correct label of x has been known.

According to Eq. (1.2), positively correlated classifiers only slightly reduce the added error, uncorrelated classifiers reduce the added error by a factor of 1/L, and negatively correlated classifiers reduce the error even further. It is clear that the smaller the correlation, the smaller the added error [Kuncheva, 2003], the same conclusion as Eq. (1.1).

2.3. *Relationship between measures of diversity and combination method*

This is a key issue. Unfortunately there is still no clear answer so far. From the experiment results with classifier combining rules by Duin [Duin, 2000] in 2000, "The best overall performance is obtained by combing both, all classifiers and all feature sets, although combining one classifier in all features also gives one overall near good result; The combined performances are not much better than the original if to choose randomized feature sets". According to Shipp and Kunchev's work [Shipp & Kuncheva, 2002], there are "very little correlation between the combination methods and the diversity measures. In fact, most of them showed independent relationships". "Since the correlation between these measures of diversity and combination methods is not very high, the question of the participation of diversity measures in designing classifier ensembles is still open."

2.4. *Measures of diversity*

A set of measures of diversity have been proposed. Kuncheva [Kuncheva, 2003] and Shipp [Shipp & Kuncheva, 2002] summarized and evaluated ten measures of diversity in 2003. Some measure the whole group of classifiers whilst others consider the classifiers on a pairwise basis and then average the results.

Table 1. Pairwise diversity measure.

	D_k correct	D_k wrong
D_i correct	A	b
D_i wrong	C	d

S. Bian & W. Wang

Pairwise measures calculate diversity values between two classifiers. After obtaining all diversity measures in a classifier poll, a confusion table is listed. Table 1.1 shows the basic idea of pairwise measures. Assume Di and Dk are two classifiers, four values a, b, c, d are defined to count the classification results of Di and Dk.

a: the number of samples that are predicted correctly by both Di and Dk

b: the number of samples that are predicted correctly by Di but not by Dk

c: the number of samples that are predicted correctly by D_k but not by D_id: the number of samples that are wrongly predicted by both D_i and D_k The total number of the test samples is: $a + b + c + d$.

In this paper, we select the Disagreement Measure (D) as main measurement. D was used by Skalak [Skalak, 1996] in 1996 and Ho [Ho, 1998] in 1998. It is the ratio between the number of observations on which one classifier is correct and the other is incorrect to the total number of observations. D is between 0, when no diversity between two learners, and 1 when they have the maximum diversity.

$$D_{i,k} = \frac{b+c}{a+b+c+d} \qquad (1.3)$$

2.5. Microarray data

DNA Microarray (DNA chips) is a recently-developed way of carrying out a single massive experiment, resulting in the analysis of a very large quantity of gene and genome data. Microarray technology has been applied to many fields of prediction and diagnosis in microbiology and medicine. However, microarray data are usually featured with a large number of attributes but few instances, which may lead to poor modelling. So far, it is still difficult to find a popular practical software platform, especially for processing microarray data. The ECS focuses on processing gene expression microarray data, addresses the problem of too many attributes with fewer samples by placing a feature subset selection (FSS) module before modelling, which identifies and eliminates redundant or irrelevant attributes. Furthermore, relevant FSS methods poll, classifier poll, and meta method poll are built in the package. This makes the program more flexible to adapt different data types and provides users more choices in setting up their systems.

3. Ensemble Classification System (ECS) Design

3.1. *ECS overview*

An ensemble system, named Ensemble Classification System (ECS), has been designed with five functional modules. ECS is extended from a open source framework – Weka. Weka is a collection of machine learning algorithms for data mining tasks coded with Java programming language [Frank, 2004].

(1) Data Process Section;
(2) Feature Selection Section;
(3) Classifiers Selection Section;
(4) Feature-classifier Pair Evaluation and Selection Section;
(5) Ensemble (combination) Section.

Figure 1 illustrates the structure of the ECS. In the ECS, raw Data are presented to the system by Data Process section. Methods are built in focus on user needs, such as data clean, partition, shuffle and format, etc. Some methods are specially suited to microarray data structure, such as data transpose. After necessary data pre-processes, the whole data are partitioned into a training set and a test set. Feature subset selection (FSS) pool provides several algorithms for user to select feature subset filters. Also classifier pool and combiner pool provide classifiers and combination classifiers (meta classifier). Diversity measure pool has algorithms for diversity computing. FSS filter applies to both training and test datasets to create two new sets. Redundant attributes of a dataset are filtered out. Classifiers selected from the classifier pool are trained by the training data, then evaluated by the test data.

The feature filters and classifiers can be treated as feature-filter classifier pairs that co-operate in processing datasets. Therefore, the total number of pairs in a system is the product of the number of the feature filters and number of the classifiers. When applying pairwise diversity measure, the total number of values can be calculated by the following equation:

$$C_n^2 = \frac{n!}{2!(n-2)!} \qquad (1.4)$$

where n is the number of total pairs. Each feature-classifier pair is evaluated individually. At the same time, the diversity among the feature-classifier pairs is measured by Eq. (1.3). After ranking the diversity values, they are saved in the system. The names of pairs that create the values are saved in a Ranking Pair List (RPL). The RPL is an important base that will be referred in pair selection phase. Before combining, pairs with higher diversity values will be selected according to RPL. The pairs that are chosen can be called "top" pairs. The top pairs then are combined in the combination section to create final results. These steps can be repeated until a satisfactory ensemble system is built up.

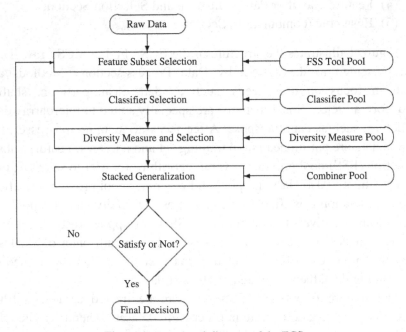

Figure 1: Functional diagram of the ECS.

3.2. *Feature subset selection*

In this study, two feature subset selection evaluations are chosen with two different search ways (Best First and Forward Selection) to produce different filters.

- CFS (correlation-based Feature Selection): CFS is a filter algorithm that ranks feature subsets according to a correlation based heuristic evaluation function. Evaluation function is 1.5:

$$Ms = \frac{k\overline{r}_{cf}}{\sqrt{k + k(k-1)\overline{r}_{ff}}} \tag{1.5}$$

Where Ms is the heuristic merit of a feature subset S containing k features, r_{cf} is the mean feature-class correlation ($f \in S$), and r_{ff} is the average feature-feature inter-correlation [Hall, 1998].

- LVF(Las Vegas Algorithm-version F): In this algorithm, probabilistic choices are made to help guide a correct solution. The worth of a subset of attributes by the level of consistency in the class values when the training instances are projected onto the subset of attributes. Consistency of any subset can never be lower than that of the full set of attributes, hence the usual practice is to use this subset evaluator in conjunction with a random or exhaustive search which looks for the smallest subset with consistency equal to that of the full set of attributes [Liu, 1996].
- Best First Search Strategy: This search is loosely based on Depth First Search with a heuristic to improve the search efficiency. The general philosophy of this strategy is to use the heuristic information to assess the "merit" latent in every candidate search avenue exposed during the search and then continue the exploration along the direction of highest merit [Dechter & Pearl, 1985].
- Forward Selection Search Strategy: This is a greedy forward search through the space of attribute subsets. It may start with no attributes or from an arbitrary point in the space, stops when the addition of any remaining attributes results in a decrease in evaluation.

3.3. Base classifiers

This study uses three kinds of classifiers: Naive Bayes, Radial Basis Function (RBF), and Support Vector Machine (SVM). The *Naive Bayes* classifier used in the ECS is developed from Flexible Bayes learning algorithm used for density kernel estimation on continuous attributes [John & Langley, 1995]. Numeric estimator precision values are chosen based on analysis of the training data. The probability density function of kernel estimation is shown in Eq. (1.6).

$$p(X = x \mid C = c) = \frac{1}{n} \sum_i g(x, \mu_i, \sigma_c) \qquad (1.6)$$

Where i ranges over the training points of attribute X in class c, and $\mu i = x i$. g is a Normal (Gaussian) distribution. The Support Vector Machine (SVM) classifier using in the ECS is developed from Sequential Minimal Optimization (SMO) [Platt, 1998]. A general non-linear SVM can be expressed as:

$$\mu = \sum_i \alpha_i y_i K(\vec{x}_i, \vec{x}) - b \qquad (1.7)$$

Where μ is the output of the SVM, K is a kernel function which measures the similarity of a stored training example $\rightarrow x_i$ to the input $\rightarrow x$, $y_i \in (-1, 1)$ is the desired output of the classifier, b is a threshold, and α_i are weights which blend the different kernels. For linear SVMs, SVM can be expressed as:

$$\mu = \vec{w} \cdot \vec{x} - b \qquad (1.8)$$

where

$$\vec{w} = \sum_i \alpha_i y_i \vec{x}_i \qquad (1.9)$$

Training a SVM requires the solution of a very large quadratic programming (QP) optimization problem. SMO breaks QP problem into a series of smallest possible QP problems and solved analytically, which reduces time consuming in training phase, especially in linear SVMs and sparse datasets [Schlkopf *et al.*, 1998]. *RBF* classifier implements a radial basis function network. It uses the K-Means clustering algorithm to provide the basis functions and learns either a logistic regression (discrete class problems) or linear regression (numeric class problems) on top of that [Hall, 2004; Bishop, 1991].

3.4. *Combination strategy*

The combination strategy in this study uses a stacking method developed from Stacked Generalisation [Wolpert, 1992]. This is a strategy to achieve a generalisation accuracy as opposed to learning accuracy. Rather than "winner-takes-all" which takes an arbitrary generaliser, final combiner combines individual learners by taking their output guesses as input components of points in a new space, and then generalizing in that new space. Figure 2 [Chan & Stolfo, 1993] illustrates the combination way employed by the ECS.

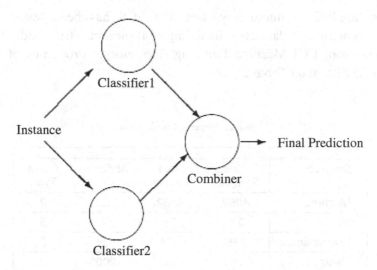

Figure 2: A combiner with two classifier.

In Fig. 2, the system consists of two generalisers – classifier1 and classifier2, and one combiner. The original learning dataset is called "level 0 space." The outputs of generalizers plus the original target value create a new instance, called "level 1 space." For example, an instance with three dimensions and one output component ($x1$, $x2$, $x3$; y) of the original dataset (level 0) passes through 2 classifiers. The two guesses from the two classifiers plus the output component generate a new instance of level 1 ($C1$ ($x1$, $x2$, $x3$), $C2(x1$, $x2$, $x3$); y). The Combiner is trained and then tested by the level 1 data. The level 1 guess also can be transformed back into a level 0, then be passed by the combiner again to create a level 2 space. Theoretically, this process can be iterated, so called "stacked generalization." In the ECS, we simply take the output values of the combiner at level 1 space as the final result. Figure 3 is a screen-shoot of the Graphic User Interface (GUI) of the ECS implemented by using Java programming language.

4. Experiments

4.1. *Experimental datasets*

Before applied to microarray data, the ECS has been tested with seveal benchmark datasets, including Mushroom, Iris and Adult datasets from UCI Machine Learning Repository. Properties of each dataset are listed on Table 2.

Table 2. Properties of datasets.

Datasets	Training Samples	Test Samples	Attributes	Class Type
Mushroom	4062	4062	22	2
Iris	75	75	4	3
Breastcancer	285	284	31	2
Gecolon	31	31	2000	2
Lymphoma	22	27	2985	2

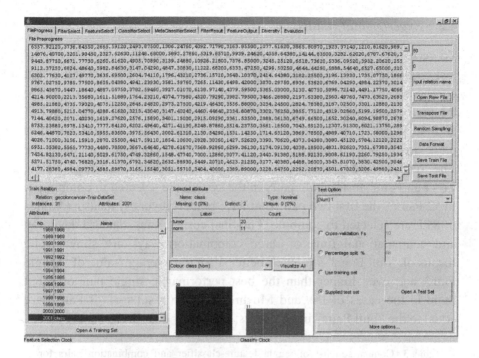

Figure 3: is a screen-shoot of the Graphic User Interface (GUI) of the ECS implemented by using Java programming language.

Mushroom data is about outcomes expressing the occurrence of edible or poisonous from 22 attributes of a mushroom. Iris dataset contains 3 classes where each class refers to a type of iris plant. Breast cancer data is about predicting benign or malignant from 30 input features based on 569 cases. Microarray data Gecolon-Cancer is from The University of Texas School of Public Health. Lymphoma cancer data is from Stanford Genomic Resources. Properties of them are also listed on Table 2. Colon cancer dataset consists of 62 samples of colon epithelial cells taken from colon-cancer patients. Each sample contains 2000 gene levels (attributes). Forty of 62 samples are cancer samples and remaining are normal samples. Lymphoma cancer data consists of 47 samples, 2985 attributes included. The target classification is to separate GC B cell-like DLCL and activate B cell-like DLCL.

4.2. *Experimental results*

To verify the ECS, some comparing tests have been done. Details of the test results for Mushroom, Iris and Breastcancer are listed in Table 3 and comparison with other researcher's work shown in Table 1.5. In Table 3, the performance of three combination ways, "Single Pair," "Full Pairs," and "2 pairs" is summarised. To "Single Pair" test, feature-classifier pairs are evaluated individually without using combination methods. The highest and lowest values are listed with the average value. To "Full Pairs" test, whole pairs in the system are selected to combining. In "2 Pairs", the first two pairs that have highest diversity measure values in RPL are selected to combine. The final correct rate, standard deviation and computational time of three combination ways are shown. The results of "Full Pairs" are much better than the average value of "Single Pair," and even higher than the best performance single pair or equal them in the cases of Iris and Mushroom. The results of "2 Pairs" are equal to the "Full Pairs" in two cases, and lower than Breastcancer case.

Table 3. Comparison list of single feature-classifier and combination pairs for different datasets.

Iris	Single Pair	Full Pairs	2 Pairs
Correct Rate (%)	80 (64–96)	97.33	97.33
STD	22.63	N/A	N/A
Time Consume (second)	10.5	30.7	13.7
Mushroom	Single Pair	Full Pairs	2 Pairs
Correct Rate (%)	90.53 (73.07–98.57)	98.57	98.57
STD	10.97	N/A	N/A
Time Consume (second)	8.6	208.2	28.4
Breastcancer	Single Pair	Full Pairs	2 Pairs
Correct Rate (%)	92.43 (89.08–95.07)	94.01	92.96
STD	2.19	N/A	N/A
Time Consume (second)	21.7	42.0	9.3

Table 4. Comparison with time consuming between the Full pairs and the Top pairs.

Datasets	Full Pairs	2 Pairs	Time Reducing (%)
Iris	30.7	13.7	55.4
Mushroom	208.2	28.4	86.4
Breastcancer	42.0	9.3	77.9

Table 5. Comparison with the correct rate between the ECS and other researcher's work.

Dataset	Other System	ECS
Mushroom	72.74	98.57
Iris	95.3	97.33
Breastcancer	96.7	94.01

Table 6. Results of microarray datasets using the ECS.

GeneColon	Single Pair	Full Pairs	2 Pairs
Correct Rate (%)	74.19 (67.74–80.65)	87.1	87.1
STD	4.88	N/A	N/A
Time Consume (second)	8.4	14.7	7.3
Lymphoma	Single Pair	Full Pairs	2 Pairs
Correct Rate (%)	69.33 (68–72)	76	80
STD	1.97	N/A	N/A
Time Consume (second)	19.6	20.1	9.6

In Table 4, time consuming reducing rates between "Full Pairs" and "2 Pairs" top pairs are listed. Time consumption is on a scale of a second. The time reducing with "2 Pairs" comparing with the "Full Pairs" in all three datasets are significant.

In Table 5, the results of the ECS are compared with other two researcher's work: John and Langley [John & Langley, 1995] applied the Flexible Bayes Classifier in Iris and Breast cancer datasets; Kerdprasop *et al.* [Kerdprasop & Kerdprasop, 2003] applied a new data partitioning algorithm in data mining using Mushroom dataset. The experiment results provided by them are considered to be one of the the best current results on the relevant datasets.   By John's paper, the total number of attributes of the Breastcancer dataset used in his test is only ten. In this study, all 30 attributes are used except the attribute ID. This probably is a reason that the correct rate for the Breast cancer data produced by the ECS is lower than John's results. Other results, of Iris and Mushroom, are better than the compared.   These comparisons indicate two things: The ECS performed as thus expected and passed the functionality verification; the performance of the ECS is better for most of the test datasets.   In the further microarray datasets experiments (results are listed in Table 6), the ECS combining with full pairs performs better than any single learner. Also it is better or equal to the full pairs combining.   On the other hand, the time consumed is much reduced.

5. Conclusion and Further Work

Our experiments indcated that ensemble is a good way to improve the classification performance. Suitable diversity measurement is helpful in either improving ensemble system performance or reducing time consuming, or both.

The ECS is a stable platform for classification. The flexibility of the ECS provides users more choices to set up their own ensemble system according to different data mining task.   Guiding with the diversity measure, the ECS has the potential to improve system performance and reduce consuming time.

Future work will include the investigation on more effective diversity measures and develop not only more accurate but also more diverse classifiers for combination in order to enhance ensemble performance.

References

Duin, Robert P.W. (2000). Experiments with Classifier Combining Rules, *In Proceeding of the First International Workshop on Multiple Classifier Systems*, LNCS 1857.

Hansen, L. and Salamon (1990). Neural network ensembles, *IEEE Transactions on Pattern Analysis and Machine Intelligence*, 12, pp. 993–1001.

Tumer,Kagan and Ghosh,Joydeep (1996). Error Correlation and Error Reduction in Ensemble Classifiers, *Connection Science*, 12, 34, 8, pp. 385–404.

Tan, Aik Choon and Gilbert, David (2003). Ensemble machine learning on gene experession data for cancer classification, *Bioinformatics*, 2, March, pp. 75–83.

Liu, Juan and Iba, Hitoshi (2001). Selecting Informative Genes with Parallel Genetic Algorithms in Tissue Classification, *Genome Informatics*, 12, pp. 14–23.

Garg,Ashutosh; Pavlovic, V. and Huang, T.S. (2002). Bayesian Networks as Ensemble of Classifiers, *ICPR (International Conference on Pattern Recognition)*, Quebec, Canada.

Wenjia, Wang and Brazier, K. (2004). Analysis of Training Injury Data Final Report.

Wolpert, David H.(1992). Stacked Generalization, *Neural Networks*, 2, May, pp. 241–259.

Shipp, C.A. and Kuncheva, Ludmila I. (2002). Relationships Between Combination Methods and Measures of Diversity in Combining Classifiers. *Information Fusion*, 3(2), pp. 135–148.

Sharkey,Amanda and Sharkey,Noe (1997). Combining diverse neural nets, *TheKnowledge Engineering Review*, 12, pp. 231–247.

Dietterich,Thomas G. (2000). Ensemble Methods in Machine Learning, Lecture note.

Rivest,Hyafil (1976). Constructing optimal decision trees is NP-complete, *Information Processing Letters*, 5, pp. 15–17.

Hall, Mark A. (1998). Correlation-based feature selection machine learning, Department of Computer Science University of Waikato, New Zealand.

Liu, Huan and Setiono,Rudy (1996). A Probabilistic Approach to Feature Selection–A Filter Solution In Machine Learning, *In Proceeding of the 13th International Conference*, Bari, Italy, July.

Dechter,Rina and Pearl, Judea (1985). Generalized best-first search strategies and the optimality of A*, *Journal of the ACM (JACM)*, July, Issue 3, 32.

Chan, Philip and Stolfo,Salvatore (1993). Experiments on Multistrategy Learning by Meta-learning, *In Proceeding of Conference on Information and Knowledge Management*, pp. 313–323.

Littlewood B., Miller D. R. (1989). Conceptual modeling of coincident failures in multiversion software, *IEEE Transactions on Software Engineering*, 15(12), pp. 1596–1614.

Kuncheva,Ludmila I. (2003). Measures of Diversity in Classifier Ensembles and Their Relationship with the Ensemble Accuracy, *Kluwer Academic Publishers*, Netherlands.

Park, Chanho and Cho, Sung-Bae (2003). Evolutionary Computation for Optimal Ensemble Classifier in Lymphoma Cancer Classification. *In Proceeding of the 14th International Symposium on Methodologies for Intelligent Systems (ISMIS) 2003*, Japan, pp. 521–530.

John, George H. and Langley, Pat (1995). Estimating Continuous Distributions in Bayesian Classifiers, *Proceedings of the Eleventh Conference on Uncertainty in Artificial Intelligence*, Morgan Kaufmann, San Mateo, pp. 338–345.

Platt, John C. (1998). Using Analytic QP and Sparseness to Speed Training of Support Vector Machines, *Neural Information Processing Systems 11*. Schlkopf, Bernhard and Burges, Chris and Smola, Alex J. (1998). Advances in Kernel Methods - Support Vector Machines, *MIT Press*, Cambridge, MA, US, 0262194163.

Quinlan, J. Ross (1993).C4.5: Programs for Machine Learning, *Morgan Kaufmann*, isbn: 1558602380.

John, George H. and Langley, Pat (1995). Estimating Continuous Distributions in Bayesian Classifiers, *The Eleventh Conference on Uncertainty in Artifical Intelligence*.

Kerdprasop, Nittaya and Kerdprasop, Kittisak (2003). Data Partitioning for Incremental Data Mining, *1st International Forum on Information and Computer Technology*, Shizuoka University, Japan.

Skalak, D. (1996).The sources of increased accuracy for two proposed boosting algorithms, *In Proceeding of American Association for Artificial Intelligence*, AAAI-96.

Ho, T. (1998). The random space method for constructing decision forests, *IEEE Transactions on Pattern Analysis and Machine Intelligence*,20(8), pp. 832–844.

Bishop, C (1991). Improving the generalisation properties of radial basis function neural networks, *Neural Computation*, 3, 4, pp. 579–588.

Hall, Mark A. (2004). http://www.cs.waikato.ac.nz/m~ hall/. Frank, Eibe (2004). http://www.cs.waikato.ac.nz/ml/weka.

Bo, Trond Hellem and Jonassen, Inge (2002). New feature subset selection procedures for classification of expression profiles, *Genome Biology*,3,4.

Cunningham, Padraig (2004). Multistrategy Ensemble Learning: Reducing Error by Combining Ensemble Learning Techniques, IEEE Transactions on Knowledge and Data Engineering, 16, 8, August.

Tsymbal, Alexey and Cunningham, Pdraig and Pechenizkiy, Mykola and Puuronen, Seppo (2003). Search Strategies for Ensemble Feature Selection in Medical Diagnostics, In Proceeding of the 16th IEEE Symposium on Computer- Based Medical Systems (CBMS'03).

CHAPTER 12

AN AUTOMATED METHOD FOR CELL PHASE IDENTIFICATION IN HIGH THROUGHPUT TIME-LAPSE SCREENS

Xiaowei Chen[1,2], Xiaobo Zhou[1,2], and Stephen T.C. Wong[1,2,*]

[1]*HCNR Center for Bioinformatics, Harvard Medical School, Boston, MA 02115*
[2]*Functional and Molecular Imaging Centery, Brigham & Women's Hospital, Boston, MA 02115*

Abstract

To better understand drug effects on cancer cells, it is important to measure cell cycle progression in individual cells as a function of time. Time-lapse fluorescence microscopy imaging provides an important method to study the dynamic cell cycle process under different conditions of perturbation. However, the assignment of a cell to a particular phase is done by visual inspection of images. To use time-lapse fluorescence microscopy for high throughput cell cycle analysis and drug screens, improved approaches that are more automated and objective are needed. In this chapter, an automated method is proposed for cell phase identification in time-lapse microscopy data. A set of twelve shape and intensity features are first extracted to describe nuclei differences in different phases based on the experience of cell biologists. We then compare the performance of different pattern recognition techniques for cell phase identification. A k-nearest neighbor classifier with a set of seven features shows the best identification result. Final identification is performed by the k nearest neighbor classifier. The accuracy of the identification is further enhanced by applying knowledge rules of cell biology.

1. Introduction

To better understand drug effects on cancer cells, it is important to measure cell cycle progression (e.g., inter phase, prophase, metaphase, and anaphase) in individual cells as a function of time. Cell cycle progress can be identified by measuring nucleus changes. Automated time-lapse fluorescence microscopy imaging provides an important method to observe and study nuclei in a dynamic fashion [6, 7, 9]. However, existing methods are inadequate to extract cell phase information of large volumes of high resolution time lapse microscopy images. Assigning a cell to a particular phase is currently done by visually inspection of images [4, 10, 17]. Such manual inspection is subject to investigator bias, cannot be easily be confirmed by other investigators, and, more importantly, cannot scale up for high throughput studies such as drug screens and cytologic profiling.

An automated system for cell phase identification would therefore have a number of advantages over current practice such as speed, objectivity, reliability, and repeatability. To the best of our knowledge, only Mackey *et al.* at University of Iowa studied mitotic cell recognition using Hidden Markov Models [5]. Their method, however, can only deal with three cell types such as dead cell, cell edge, and dividing cell. No other work has successfully been done on automated cell phase identification. This motivates us to develop new methods for quantitatively describing cell nuclei patterns in different phases from massive volumes of dynamic cellular images. In this study, twelve features are extracted to describe the shape and intensity difference of cell nuclei in different mitotic phases. We then compare the performance of different feature selection, feature reduction techniques and classification techniques. We use sequential forward selection (SFS) [12] for feature selection, linear discriminant analysis (LDA) for feature reduction, k nearest neighbor (KNN) classifier [8] and maximum-likelihood (ML) classifier [16] for cell phase identification. Since we are dealing with time series data some temporal information can be used to refine the classifier output. In this study we apply a set of knowledge driven rules to compensate for certain phase identification errors from syntactical classifiers.

Our automated cell phase identification method included three parts: nuclei segment, nuclei tracking, and cell phase identification. The procedure is schematically summarized in Fig. 1 and discussed in the following sections.

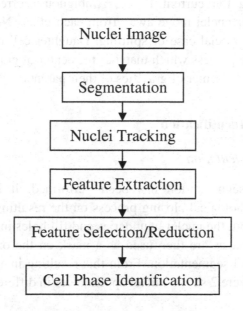

Figure 1: Summary of the method for automated cell phase identification.

2. Nuclei Segmentation and Tracking

Nuclei segmentation, which segments individual nuclei from their background, is very critical for cell phase identification. In time-lapse fluorescence microscopy images, nuclei are bright objects protruding out from a relatively uniform dark background. Thus, they can be segmented by histogram thresholding. In this work, the ISODATA algorithm was used to perform image thresholding [15]. This algorithm correctly segments most isolated nuclei, but it is unable to segment touching nuclei. To handle the case of touching nuclei, a watershed algorithm was used [1, 14]. Watershed algorithms always cause nuclei be to over-segmented. To solve this problem, a size and shape based merging processing [3] is used to put over-segmented nuclei pieces together.

We also developed a size and location based method for nuclei tracking, which can successfully deal with nuclei changes during cell mitosis and cell division [3]. In this method, a match processing is used to set up the correspondence between nuclei in the current frame and the frame following the current frame. Ambiguous correspondences are solved only after nuclei move away from each other. Nuclei division is considered as a special case of splitting. Daughter cell nuclei are found and verified by a process which matches the center of gravity of daughter cell nuclei with the center of gravities of their parents.

3.   Cell Phase Identification

3.1.   *Feature calculation*

After nucleus segmentation has been performed, it is necessary to perform a morphological closing process on the resulting binary images in order to smooth the nuclei boundaries and fill holes insides nuclei [2]. These binary images are then used as a mask on the original image to arrive at the final segmentation. From this resulting image, features can be extracted. Figure 2 shows nuclei appearances in different phases.

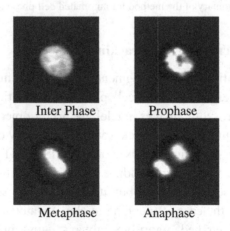

Figure 2: Nucleus appearances in different phase.

Table 1. List of features used for cell phase identification.

Features	Name	Description
1	Max Intensity	Max, min, average, and stand deviation of gray level within a cell nucleus
2	Min Intensity	
3	Mean	
4	Stand Deviation	
5	Major Axis	the length of a cell nucleus
6	Minor Axis	the breadth of a cell nucleus
7	Elongation	Major Axis / Minor Axis
8	Area	Number of pixels within a cell nucleus
9	Perimeter	The total length of edges of a cell nucleus
10	Compactness	Perimeter / Area
11	Convex Perimeter	perimeter of the convex hull
12	Roughness	Convex Perimeter / Perimeter

Once the individual nuclei have been detected by the segmentation procedure, it is then necessary to identify cell phase. Cell phase are identified by comparing their nuclei shape and intensity. To identify the shape and intensity differences between different cell phases, a set of twelve features were extracted based on the experience of expert cell biologists. Table 1 lists these features. Since the feature values have completely different ranges, an objective scaling of features was achieved by calculating z-scores [11] given by

$$z_{ij} = \frac{x_{ij} - \overline{x}_j}{s_j} \qquad (1)$$

where x_{ij} represents j-th feature of the i-th nucleus and $\bar{m}_j$ is the mean value of all n cells for feature j, and s_j is the mean absolute deviation [11], i.e., $s_j = \dfrac{1}{n}\sum_{i=1}^{n}|x_{ij} - \bar{x}_j|$.

3.2. *Identifying cell phase*

In pattern recognition, feature selection and feature reduction are two very important steps to improve recognition accuracy and save computational cost. Next we will introduce two methods, namely: feature selection using sequential forward feature selection, and feature reduction using linear discriminant analysis. In cell phase identification, we will consider two classical classifiers, namely nonparameter classifier k-nearest neighbor (KNN) and parameter classifier maximum likelihood (ML) method.

A. *Feature Subset Selection*

In this work, the sequential forward selection (SFS) method was used to for feature selection. SFS is a bottom-up search procedure where features are added to the current feature set one by one. At each stage, only one feature is selected from the remaining features to add to the current feature set. If the added feature set yields a better classification error rate than adding any other single feature, it is used. The optimal feature set has been found when no feature can be added to the current feature set which reduces the classification error rate. Note that SFS is a suboptimal feature selection method, and it could fail when two features are good together but not separately, so we also studied feature reduction such as linear discriminate analysis and compare the performance of the different feature selection methods. The performance of a classifier can provide a criterion to evaluate the discrimination power of the features for the feature subset selection. In this work, we choose a KNN classifier for its simplicity and flexibility.

B. Linear Discrimininant Analysis

Feature selection and transformation such as principle component analysis and linear discriminant analysis (LDA) are main ways for feature reduction in data analysis. Since LDA processes the data matrix of training samples to find a linear transformation that enhances discrimination of the different categories, it is more suitable for feature reduction in pattern recognition. Recently, there have been attempts at using alternative metrics such as [17] to select the number of components. After the feature extraction and normalization, we still denote the feature vector as z_{ij} $i = 1,...,$n$, j = 1,...,m$ where n is the total number of samples, and m is the number of feature. Let C denote total number of phases, say four phases in this chapter, and Ω_c denote the set of all samples in class C. Denote

$$\overline{z}_{cj} = \frac{1}{|\Omega_c|} \sum_{z_{ij} \in \Omega_c} z_{ij} \qquad (2)$$

where $|\Omega_c|$ is the number of samples in class C. $\mathbf{z}_i = [z_{i1}, z_{i2},..., z_{im}]$ and $\overline{\mathbf{z}}_c = [\overline{z}_{c1}, \overline{z}_{c2},..., \overline{z}_{cm}]$. Next we define within-class scatter matrix

$$W = \frac{1}{n} \sum_{i=1}^{n} \sum_{z_{ij} \in \Omega_c} (\mathbf{z}_i - \overline{\mathbf{z}}_c)^T (\mathbf{z}_i - \overline{\mathbf{z}}_c), \qquad (3)$$

and the total scatter matrix is defined as

$$T = \frac{1}{n} \sum_{i=1}^{n} (\mathbf{z}_i - \overline{\mathbf{z}})^T (\mathbf{z}_i - \overline{\mathbf{z}}) \qquad (4)$$

where $\overline{\mathbf{z}} = [\overline{z}_1, \overline{z}_2,..., \overline{z}_m]$ and $\overline{z}_j = \frac{1}{n} \sum_{i=1}^{n} z_{ij}$. We then perform orthogonal decomposition. Denote the eigenvector matrix and diagonal eigenvalue matrix of matrix $W^{-1}T$ as Φ and Λ. Then we have

$$W^{-1}T\Phi = \Phi_{m \times m} \Lambda_{m \times m} \qquad (5)$$

where $\Lambda = diag(\lambda_1, \lambda_2, ..., \lambda_m)$. Without loss of generality, the eigenvalues are sorted in descending order. Usually, we select the biggest eigenvalues that the sum of these eigenvalues is 90% of sum of total eigenvalues. Assume such biggest eigenvalues are $\lambda_1, \lambda_2, ..., \lambda_L$ $(L \le m)$. One can also select L by experience. Denote the matrix composed of the corresponding eigenvectors to be $\hat{\Phi}_{m \times L}$. Then the data $z_i = (z_{i1}, z_{i2}, ..., z_{im})$ is transformed as following $u_i = z_i \hat{\Phi}$ with $u_{ij} = (u_{i1}, u_{i2}, ..., u_{iL})$ where $i = 1, ..., n$.

C. K-Nearest Neighbor Classifier

KNN classifier is a non-parametric classifier, which is easy to implement. In KNN, each cell is represented as a vector in a m-dimension feature space. The distance $d_E(x, y)$ between a cell $x = (x_1, ..., x_m)^T$ and a cell $y = (y_1, ..., y_m)^T$ is defined by the Euclidian distance. A training set T is used to determine the class of a previously unseen nucleus. First, the classifier calculated the distances between an unseen nucleus and all nuclei in the training set. Next, the K cells in the training set which are the closest to cell x are selected. The phase of cell x is determined to be the phase of the most common cell type in the K nearest neighbors.

D. Maximum-Likelihood Classifier

The ML classifier is a parametric classifier that relies on the second-order statistics of a Gaussian probability density function (pdf) model for each class. It is often used as a reference for classifier comparison because. If the class pdf's are indeed Gaussian, it is the optimal classifier. The basic discriminate function for each class is

$$g_c(\mathbf{u}) = p(\mathbf{u} \mid w_c)p(w_c) = \frac{p(w_c)}{(2\pi)^{m/2}\Sigma_c^{1/2}} \exp\left\{-\frac{1}{2}(\mathbf{u} - \bar{\mathbf{u}}_c)\Sigma_c^{-1}(\mathbf{u} - \bar{\mathbf{u}}_c)^T\right\} \quad (6)$$

where $\mathbf{u}$ is the data vector, $\bar{\mathbf{u}}_c$ is the mean vector of class c, and Σ_c is the covariance matrix of class c, $\Sigma_c = \frac{1}{|\Omega_c|} \sum_{z_{ij} \in \Omega_c} (\mathbf{u}_i - \bar{\mathbf{u}}_c)^T (\mathbf{u}_i - \bar{\mathbf{u}}_c)$. The values in the mean vector, and the covariance matrix, are estimated from the training data by the unbiased estimators. If the a priori probabilities $p(w_c)$ are assumed to be equal, this leaves only $p(\mathbf{u}|w_c)$ to be calculated for each sample during classification. The discriminate $g_c(\mathbf{u})$ is calculated for each class and the class with the highest value is selected for the final classification map.

3.3. *Correcting cell phase identification errors*

One observation is that even though cell cycle process can be affected by adding different drug treatments, the change is confined by biological constraints. Thus, by applying knowledge driven heuristic rules some tracking errors may be corrected. Our automated method uses the following three known biological rules:

- Phase progression rule: Once a cell enters a defined cell-cycle phase, it cannot go back to its previous phase, i.e., it passes the point of no return.
- Phase continuation rule: Cells cannot skip one phase and enter the phase following the one it skipped. In some case, a cell may stay in prophase for less than 15 minutes, and this will result in missing a phase in the cell sequence. However, cells cannot jump from metaphase to inter-phase or from anaphase to metaphase.
- Phase timing rule: The time period that a cell stays in a phase also follows certain general rules. Cells will usually stay in prophase for no more than 45 minutes, metaphase for around 1 hour in untreated sequences, and anaphase under 1 hour. The time that a cell stays in inter-phase is usually more than 20 hours and is much longer than the time it stays in various mitotic phases. In time-lapse sequences of drug-treated cell population, certain cells can stay in metaphase for a little longer time or remain there to the end of sequence.

To illustrate how to apply these heuristic rules to correct cell phase identification errors by the KNN classifier, four examples are provided in Fig. 3, and each of them shows a portion of nucleus sequences. Bold font marks the places where the errors happen. In Case A, the inter phase cell is misclassified as prophase cell four times, and these errors can be detected and corrected by applying the phase progression rule. In Case B, the inter-phase cell is misclassified as anaphase cell three times, and these errors can be detected and corrected by applying the phase continuation rule. Cells in certain periods of prophase look similar to cells arrested in metaphase (Once cells enter metaphase, they stay there and will not divide further, which is called 'arrested' in metaphase.) This will cause some misclassification between prophase cells and arrested metaphase cells. From a biology point of view, prophase begins when cells try to align their chromosomes and ends when the chromosomes are aligned. When this alignment process cannot be finished because of the influence of drugs, cells are arrested. Thus, cells arrested in metaphase are essentially the same as cells in the middle of prophase. To deal with these kinds of errors, metaphase is further divided into to normal metaphase and arrested metaphase. In Case C, in the first seven frames the metaphase cell is misclassified as prophase cell, and these errors can also be detected and corrected by applying the phase timing rule. The remaining one error can be detected and corrected by applying the phase continuation rule. In Case D, one prophase cell is misclassified as arrested metaphase cell. This error can be corrected by applying the phase timing rule.

A: 1,1,1,1,1,1,1,1,1,**2**,1,1,**2**,1,1,1,1,1,1,**2**,1,1,1,1,1,1,**2**,1,1,1,1,1,1

B: 1,1,1,1,1,1,1,1,1,**4**,1,1,1,1,**4**,1,1,1,1,1,1,**4**,1,1,1,1,1,1,1,1,1,1,1

C: **2,2,2,2,2,2,2**,5,5,5,5,5,5,5,5,5,5,**2**,5,5,5,5,5,5,5,5,5,5,5,5,5,5,5

D: 1,1,1,1,1,1,1,1,**2,5**,3,3,3,3,3,4,4,4,1,1,1,1,1,1,1,1,1,1,1,1,1,1,1

Figure 3: Cases A-D show portions of cell sequences. 1 stands for inter phase, 2 prophase, 3 metaphase, 4 anaphase, and 5 arrested metaphase.

4. Experimental Results

We used nucleus images from four nucleus sequences to test the efficiency of the proposed method. Each sequence consists of ninety-six frames over a period of 24 hours. The sequences were recorded at a spatial resolution of 672*512, and a temporal resolution of one image per 15 minutes with a time-lapse fluorescence microscopy. The cancer cell line used is HeLa cells in our experiments. Two types of sequences were used to denote drug treated and untreated cells. Cell cycle progress was affected by treatment of drug compounds, and some or all of the cells in the treated sequences were arrested in metaphase, whereas cell cycle progress in the untreated sequences was not affected. Cells without drug treatment will usually undergo one division within this period.

The nucleus set used to training the feature selection method includes 100 nuclei for each phase which results a training set of 400 nuclei. The 400 nuclei were evenly divided into five disjoint subsets. Selection performance was evaluated by a five-fold cross validation in five individual tests with 4/5 of the initial data serving as the training set for the selection algorithm. The remaining 1/5th of the data served as the test set.

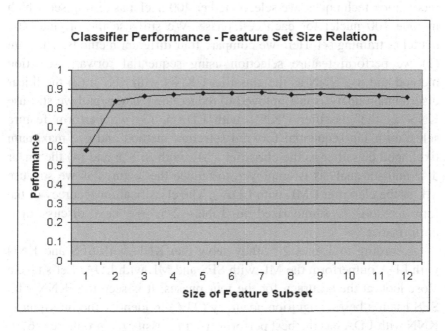

Figure 4: Result of feature subset selection.

In exhaustive experiments, a six nearest-neighbor rule delivered the most reliable results for the different selection strategies. Figure 4 shows the classifier performance changes, which is defined as the fraction between the number of nuclei correctly identified and the total number of nuclei, against the size of feature subset. The best performances are achieved with a subset size of seven. Adding the remaining 5 features cause the selection percentages to reduce. The features in the optimal feature set and the order they were selected by the SFS method as follows: Perimeter, Stand Deviation, Compactness, Max Intensity, Major Axis, Mean, and Minor Axis. These seven features were then used for cellular phase identification. As we mentions before that SFS could fail when two features are good together but not separately. Since we only used 12 features in this study, feature selection is a necessary but not critical. Even it fails, for examples, we can just use all features to perform classification, and the performance is slightly worse than the case based on SFS, see Fig. 4.

In order to test the performance combination of different pattern recognition techniques, we select another 400 nuclei as testing set, which include 100 nuclei for each cell phase. We still use the original 400 nuclei as training set. Here we compare four different methods. They are (1) we perform feature selection using sequential forward selection method and use KNN as the classifier ('KNN with SFS'); (2) the linear discriminant analysis is employed to reduce the features and we still use KNN as the classifier ('KNN with LDA'); (3) we perform feature selection using sequential forward selection method and use maximum likelihood classifier as the classifier ('ML with SFS'); and (4) the linear discriminant analysis is employed to reduce the features as we still use ML as the classifier ('ML with LDA'). The classification accuracy of the four methods is summarized in Tables 2-5. We next discuss their performance.

According to Tables 2-5, they show that KNN with SFS and KNN with LDA outperform the ML with SFS and ML with LDA. Let's take a close look at the accuracy for the four phases. It is seen that KNN with SFS has the best recognition accuracy (93%) to identify the inter-phase; KNN with LDA has the best performance to classify the prophase (76%); KNN with SFS (94%) and ML with SFS (94%) slightly outperform the

KNN with LDA (93%) and ML with LDA (92%) when to identify the metaphase; however the KNN with LDA (100%) and ML with LDA (100%) outperform KNN with SFS (97%) and ML with SFS (98%) when to identify the anaphase. We can also see that each method has a poor performance to separate prophase nuclei with metaphase nuclei because of the similarity between them. In summary, KNN with SFS and KNN with LDA outperform other two, and they have a similar performance. Since in untreated cell sequence most of the cells are in inter phase we choose the KNN with SFS combination which has better performance that the KNN with LDA combination to identify inter phase nuclei. Next we will apply the KNN with SFS to study more nuclei.

Table 2. Cell phase (CP) identification accuracy using KNN with SFS (Asg: Assigned, I: Interphase, P: Prophase, M: Metaphase, A: Anaphase, /: Unknown. The same with Table 3, 4, 5, 6, and 7.).

True \ Asg	I	P	M	A	/	Accuracy
I (100)	93	0	0	2	5	93%
P (100)	20	70	2	1	7	70%
M (100)	0	3	94	1	2	94%
A (100)	0	0	2	97	1	97%

Table 3. CP identification accuracy using KNN with LDA.

True \ Asg	I	P	M	A	/	Accuracy
I (100)	88	1	0	7	4	88%
P (100)	15	76	0	3	6	76%
M (100)	0	2	93	3	2	93%
A (100)	0	0	0	100	0	100%

X. Chen, X. Zhou & S. T. C. Wong

Table 4. CP identification accuracy using ML with SFS.

Asg True	I	P	M	A	Accuracy
I (100)	89	6	0	5	89%
P (100)	2	54	36	7	54%
M (100)	0	1	94	5	94%
A (100)	0	0	2	98	98%

Table 5. CP identification accuracy using ML with LDA.

Asg True	I	P	M	A	Accuracy
I (100)	82	11	0	7	82%
P (100)	15	66	13	6	66%
M (100)	0	2	92	6	92%
A (100)	0	0	0	100	100%

A total of 80 nuclei were selected from the four sequences. Each nucleus was tracked for 12.5 hours. Thus 50 images were taken for each nucleus. During this time, these 80 cells either divided or arrested in metaphase. This process generated a test set with 4,000 nuclei. The cell phase identification experiments were performed on this test set. The cell identification was performed with a 6-Nearest Neighbor classifier based on the 7 derived features. The training set for the classifier consisted of the 400 nuclei used for feature selection. Table 6 shows the experiment results, which are compared with human analyzing result.

Table 6. Cell phase identification results.

Asg / True	I	P	M	A	/	Accuracy
I (2970)	2763	1	22	2	2	99%
P (47)	3	24	19	1	0	51.1%
M (952)	23	125	792	11	1	83.2%
A (209)	2	0	32	175	0	83.7%

Table 7. Cell phase identification results after applying heuristic rules.

Asg / True	I	P	M	A	/	Accuracy
I (2970)	2785	0	4	1	0	99.8%
P (47)	3	39	4	1	0	83%
M (952)	7	31	909	5	0	95.5%
A (209)	1	0	8	200	0	95.7%

The classifier correctly identified nearly all (99%) interphase cells. For cells in metaphase and anaphase, the accuracy of the classifier algorithm was about 83% for both of them. However, only 51.1% of cells in prophase were correctly identified. The classifier made a number of mistakes on separating metaphase cells from prophase cells. 40.4% of prophase cells are assigned as metaphase cells, and 13.1% of metaphase cells are assigned as prophase cells. Table 7 lists the phase identification results achieved by apply the knowledge driven heuristic rules to the classifier outputs. Note that most of the phase identification errors between prophase and metaphase are corrected. Only 8.5% of prophase cells are assigned as metaphase cells and 3.4% of metaphase cells are assigned as metaphase cells. The phase identification correct rate increases 0.8% for inter phase cells, 31.9% for prophase cells, 12.3% for metaphase cells, and 12% for anaphase cells.

5. Conclusion

Time-lapse cell imaging is becoming an important tool in understanding quantitative behavior of cell cycle phases and in performing high throughput cancer drug screens. However, the lack of automated data analysis method is the rate limiting factor of this new modality. In this chapter, we described a successful approach for automated cell phase identification in time-lapse fluorescence microscopy data. By combining a k nearest neighbor classifier with a set of seven features we achieved the following correct identification performance on the three cell cycle phases: inter phase 99.8%, prophase 83%, metaphase 95.5%, and anaphase 95.7% after applying knowledge rules of domain experts. For large scale cell cycle study, biologists are more interested in the statistical distribution of each phase with respect to the dwelling time. According to biologists, an automate system with more than 80% phase identification accuracy will be sufficient for such study. Thus, the result of our approach can be used directly by biologists and paves the way for large scale data mining in high throughput drug screens.

Acknowledgement

The authors would like to acknowledge the productive collaboration with their biology collaborators in this research effort, and, in particular, Dr. Randy King and Dr. Susan Hyman, of the Department of Cell Biology and ICCB (Institute of Chemistry and Cell Biology), Harvard Medical School, Boston, MA, USA. The datasets used in this study is generated in Dr. King's lab. This research is funded by a R01 LM008696 to STCW.

References

[1] Bleau A. and Leon J.L., "Watershed-based segmentation and region merging," Computer Vision and Image Understanding, 77, 3 (2000) 317-370.
[2] Chen S. and Haralick M. "Recursive erosion, dilation opening and closing transform," IEEE Transactions on image processing, 4, (1995) 335-345.

[3] Chen X. Zhou X., and S.T.C. Wong "Automated segmentation, classification, and tracking cancer cell nuclei in time-lapse microscopy," Submitted to IEEE Transactions on Biomedical Engineering.

[4] Endlich B., Radford I.R., Forrester H.B., and Dewey W.C. "Computerized video time-lapse microscopy studies of ionizing radiation-induced rapid-interphase and mitosis-related apoptosis in lymphoid cells," Radiat Res. 153, 1, (2000) 36-48

[5] Gallardo G.M., Yang F., Ianzini F., Mackey M., Sonka M., "Mitotic cell recognition with hidden Markov models," Proceedings of the SPIE, 5367(2004) 661-668.

[6] Haraguchi T., Ding D.Q., Yamamoto A., Kaneda T., Koujin T., Hiraoka Y., "Multiple-color fluorescence imaging of chromosomes and microtubules in living cells," Cell Struct Funct 24(1999) 291-298

[7] Hiraoka Y and Haraguchi T, "Fluorescence imaging of mammalian living cells," Chromosome Res 4, (1996) 173-176.

[8] Jain A.K., Duin R., and Mao J., "Statistical pattern recognition: A review," IEEE Transactions on Pattern Analysis and Machine Intelligence, 22, 1 (2000) 4-37.

[9] Kanda T., Sullivan K. F. and Wahl G. M, "Histone-GFP fusion protein enables sensitive analysis of chromosome dynamics in living mammalian cells," Current Biology 8, (1998) 377-85.

[10] Karlsson C., Katich S., Hagting A., Hoffmann I., and Pines J. "Cdc25B and Cdc25C differ markedly in their properties as initiators of mitosis," Journal Cell Biology 146(1999) 573-584

[11] Kaufman L. and Fousseeuw P. J. Finding Groups in Data: An Introduction to Cluster Analysis. New York Wiley. (1990).

[12] Kittler J., "Feature selection algorithm," in Pattern Recognition and Signal Processing (Chen C.H. ed.), Sijthoff & Noordhoof, Alphen aan den Rijn, The Netherlands (1978) 41-60.

[13] Lobo A., "Image segmentation and discriminant analysis for the identification of land cover units in ecology," IEEE trans. on Geosc. Remote Sensing, 35, 5 (1997) 1136-1145.

[14] Norberto M., Andres S., Carlos Ortiz S. Juan Jose V., Francisco P., and Jose Miguel G., "Applying watershed algorithms to the segmentation of clustered nuclei," Cytometry 28 (1997) 289-297.

[15] Otsu N. "A threshold selection method from gray level histogram," IEEE Transactions on System man Cybernetics 8 (1978) 62-66.

[16] Paola J. D. and Schowengerdt R. A., "A detailed comparison of backpropagation neural network and maximum-likelihood classifiers for urban land use classification," IEEE Trans. on Geosc. and Remote Sensing, 33, 4 (1995).

[17] Windoffer R. and Leube R.E., "Detection of cytokeratin dynamics by time-lapse fluorescence microscopy in living cells," Journal Cell Science 112 (1999) 4521-4534

CHAPTER 13

INFERENCE OF TRANSCRIPTIONAL REGULATORY NETWORKS BASED ON CANCER MICROARRAY DATA

Xiaobo Zhou and Stephen TC Wong

HCNR Center for Bioinformatics, Brigham and Women's Hospital, Harvard Medical School

Abstract

Construction of transcriptional regulatory networks using microarray techniques is a prevalent method in cancer research. In this article, we reviewed the existing approaches of constructing gene regulatory subnetworks and then presented our own approaches based on a multinomial probit model and a Monte Carlo approach.

1. Introduction

Oligonucleotide DNA array and cDNA array [20] are two well-known microarray technologies. Data from microarray experiments are usually in the form of large matrices of gene expression levels under different experimental conditions, and frequently there are missing values which must be estimated [5, 54, 61]. Since the microarray data are usually highly noisy, normalization on these data must be performed [11, 12, 55, 60]. After the preprocessing steps, functional analysis is then applied to the normalized microarray data to explore the gene pathways and gene interactions.

275

A central goal of molecular biology research is to discover the regulatory mechanisms governing the expression of genes in the cell [46]. The complex functions of a living cell are carried out through the concerted activity of many genes and gene products. The gene microarray makes it possible for us to understand gene regulation and interactions. Genome sequences specify the gene expression that produces living cells, but how cells control global gene expression remains unclear. The expression of a gene is controlled by many mechanisms. A key factor in these mechanisms is mRNA transcription regulation by various proteins, known as transcriptional factors (TFs), binding to specific sites in the promoter region of a gene that activate or inhibit transcription. Recently, there are several attempts to relate promoter sequence data and expression data [9, 38, 40, 52], as well as some interesting works on inferring network motifs [34, 36, 47] in the transcriptional regulation networks and module networks under the Bayesian framework. The network motifs are defined as patterns of interconnections that recur in many different parts of a network at high frequencies. Each network motif has a specific function in determining gene expressions. These motifs correspond to different subnetworks [13] in biological regulatory networks. The challenge is to find the network motifs based on large amounts of gene microarray and sequence data generated [34, 36].

In [39], support vector machine was employed to predict regulatory networks based on identification transcription factors from gene expression data. In Refs. 45 and 46 a probabilistic framework is proposed that models the process by which transcriptional binding explains the mRNA expression of different genes. This model unifies two important components: the prediction of gene regulation events from sequence motifs in the gene's promoter region and the prediction of mRNA expression from a combination of gene regulation events. This model is called module networks, and it is constructed under the framework of the Bayesian networks. In Ref. 51, a similar method to constructing gene networks from expression profiles by combining Bayesian network with promoter element detection was proposed.

There have been a number of approaches to modeling gene regulatory networks, including linear models [21, 50], Bayesian networks [13, 15,

16, 17, 23, 37], neural networks [57], differential equations [10, 35], and models incorporating stochastic components at the molecular level [1, 24, 49]. Among the modeling systems received the most attention include the Bayesian networks and the Boolean networks originally introduced by Friedman [15] and Kauffman [19, 30, 31], respectively. There are a number of works applying Boolean networks (BNs) to genomic analysis, such as [1, 2, 22, 27, 35, 48, 59, 62, 63]. An important step in constructing these networks involves finding genes that have the strongest influence on the target gene. We will focus on this problem in this chapter. Many works exist on regulatory network construction, but very few of them concern the important problem of subnetwork construction.

The remainder of this chapter is organized as follows. Section 2 reviews existing subnetworks and regulatory network construction methods. In Section 3, we present gene prediction and network construction based on probit regression with Bayesian gene selection. In Section 4, we present network construction based on clustering and Markov chain and Monte Carlo (MCMC) predictor design. Section 5 presents the conclusion.

2. Subnetworks and Transcriptional Regulatory Networks Inference

There are three common ways to start the construction of regulatory networks: starting with certain optimal criteria and objective functions; starting with graph theory, and starting with clustering genes. In this section, we present four existing approaches to constructing subnetworks and networks: inferring subnetwork based on Z-score, graph theory, Bayesian score, and, finally, based on integrated expression and sequence data.

2.1. *Inferring subnetworks using z-score*

In Ref. 28, the authors proposed to infer subnetworks based on z-score. Let p-values p_i represent the significance of expression change for each gene i, and each p_i is then converted to a z-score $z_i = \Phi^{-1}(1 - p_i)$, where

Φ^{-1} is the inverse normal CDF. An aggregate z-score z_A for an entire subnetwork A of k genes is defined as

$$z_A \triangleq \frac{1}{\sqrt{k}} \sum_{i \in A} z_i \tag{1}$$

Subnetworks of all sizes are comparable under this scoring definition. A high z_A indicates a biologically active subnetwork. Given k genes, we have $\binom{n}{k}$ subnets, so it is necessary to normalize the corresponding values defined in (1): Randomly sample gene sets of size k using a Monte Carlo approach, then compute their scores z_A, and, finally, estimate mean μ_k and standard deviation σ_k for each k. Using these estimates, the corrected subnet score s_A is defined as:

$$s_A \triangleq \frac{z_A - \mu_A}{\sigma_k} \tag{2}$$

The scoring system can be extended to multiple conditions. In this case, we have a matrix of p-values and corresponding z-scores. Given a subnetwork A, generate m different aggregate score $(z_{A1}, z_{A2}, ..., z_{Am})$, one for each condition. Then sort them from the highest to the lowest: $(z_{A(1)}, z_{A(2)}, ..., z_{A(m)})$, and compute the significance $r_{A(j)}$ of the jth highest score using a binomial order statistic. The maximum of these new scores is given by $r_A^{max} = \max_{1 \leq j \leq m} (r_A(j))$. The r_A^{max} is the final corrected score s_A of subnetwork A. Lastly, simulated annealing is employed to search for the subnetwork with the highest score.

2.2. Inferring subnetworks based on graph theory

Dynamic systems are especially interested in genetic networks inference. In Ref. 7, the authors proposed a duplication growth model of gene expression networks to construct dynamic networks. The time dependence model used in this work is given by

$$A(t) = \Lambda_1 A(t-1) + A(t-1) A^T(t-1) \Lambda_2 \tag{3}$$

where $A(t)$ is a matrix of the gene expression profiles, i.e., $A(t) = (a(2), ..., a(t))$ where $a(t)$ is a vector representing the expression levels of all genes in genome at time t. The matrix Λ_1 gives the influence of the elements of the jth gene on the production of the ith gene. The two matrices Λ_1 and Λ_2 generated by this data analysis are components of the weighted connectivity matrix of a graph of interactions between gene expression levels. The absolute values of the matrix elements are set equal to 1 if they are above a certain threshold, and 0 otherwise.

Gene duplication [7] is a mechanism for network growth. The basic idea of this method is described in Fig. 1: Figure 1(a) illustrates that a new node is created by duplicating the connectivity of its parents; the partial duplication model in Fig. 1(b) consists of duplication plus random removal of edges from the daughter node; Figure 1(c) depicts that a second model duplication plus preferential rewiring involves duplication followed by random rewiring of one of the edges in the network. A random network seed and a network seed with a high clustering coefficient were used as small initial seed networks. These small networks then grow based on the above criteria. The shortage of this method is that it is difficult to predict what kind of properties the final subnetworks would have.

2.3. Inferring subnetworks based on Bayesian networks

Here we introduce subnetworks networks based on some optimal criteria. The basic idea of this kind of approach is to define a score for all subnetworks with or without number of genes being a fixed number. Then heuristics search algorithm can be employed to find the subnetwork with optimal score. In this subsection, we will review the Bayesian score based method [13].

Consider a finite set $\mathbf{X} = \{x_1, x_2, ..., x_n\}$ of discrete random variables where each variable X_1 may take on values from a finite set. A Bayesian network is an annotated directed acyclic graph that encodes a joint probability distribution over $\mathbf{X}$. Formally, a Bayesian network [25] for $\mathbf{X}$ is a pair $B = < G, \Theta >$. The first component, namely G, is a directed acyclic graph whose vertices correspond to the random variables

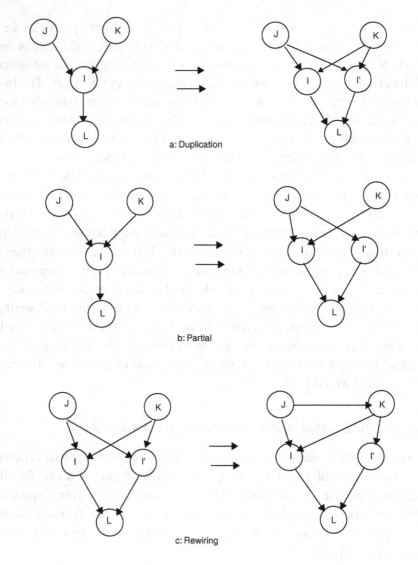

Figure 1: Network growth through gene duplication [7]: (a) a new node is created by duplicating the connectivity of the parent; (b) the partial duplication model where node I is duplicated to I' and maintain the original connection; and (c) a rewiring process where edge I-J is rewired to become I-K.

$x_1, x_2, ..., x_n$. The graph encodes the following set of conditional independence assumptions: each variable x_i is independent of its non-descendants given its parents in G. The second component of the pair, Θ, represents probability distribution for each node: $\theta_i = \theta_{i|pa(x_i)} \triangleq p(u_i | pa(x_i))$ for each possible value u_i of x_i, and $pa(x_i)$ of $Pa(x_i)$. Here $Pa(x_i)$ denotes the set of parents of x_i in G and $pa(x_i)$ is a particular instance of the parents. If more than one graph is discussed, then we use $Pa^G(x_i)$ to specify x_i's parents in graph G. A Bayesian network B specifies a unique joint probability distribution over **X** given by

$$p(x_1, x_2, ..., x_n) \triangleq \prod_{i=1}^{n} p(x_i | Pa^G(x_i)) \qquad (4)$$

The problem of learning a Bayesian network can be stated as follows. Given a training set $O = \{\underline{x}_1, \underline{x}_2, ..., \underline{x}_m\}$ of instances of **X**, find a network B that best matches O. Here $\underline{x}_i \triangleq [x_{i1}, x_{i2}, ..., x_{in}]$ and m is the number of observations. The common approach to this problem is to introduce a scoring function that evaluates each network with respect to the training data, and then to search for the best network with the highest score. The scoring function most commonly used to learn about Bayesian networks is the Bayesian score metric $P(G|O)$ [25]. According to the Bayes' rule,

$$P(G|O) = \frac{P(O|G)P(G)}{\sum_{G'} P(O|G')P(G')} \propto P(O|G)P(G) \qquad (5)$$

where $P(G)$ is *the priori* probability of the graph structure [17, 25], and $P(O|G)$ is the marginal likelihood given by $P(O|G) = \int p(O|G, \Theta)p(\Theta)d\Theta$. If O is complete and $P(G)$ satisfies parameter independence according to (4), then the marginal likelihood decomposes into a product of terms given by:

$$P(O|G) = \prod_{i=1}^{n} score(x_i, Pa^G(x_i) | O) \qquad (6)$$

where $score(x_i, Pa^G(x_i) | O)$ has a simple closed-form solution [17, 25] for standard priors such as Dirichlet or Wishart priors.

In Bayesian model averaging, we estimate the probability $P(x \rightarrow y)$ of a feature from note x to node y using the Monte Carlo method. This can be done by searching the directed acyclic graphs (DAG) using the improving Markov chain Monte Carlo (MCMC) model search method [18, 43]. For example, the Markov chain length is chosen as 10,000 after the 10,000 burn-in period, then $P(x \rightarrow y)$ is estimated by:

$$P(x \rightarrow y) \approx \sum_{j=10001}^{20000} P(G_j | O) \delta(x, y, G_j) \tag{7}$$

where $\delta(x, y, G_j)$ is 1 if G_j contains feature $(x \rightarrow y)$, and is 0 otherwise. In this way, we can compute the posterior probability of all possible edges. Based on a pre-defined threshold α, if $P(x \rightarrow y) > \alpha$, then there exists a feature $(x \rightarrow y)$. The application of Bayesian networks in modeling gene regulatory networks can be found in [13, 15, 16, 25, 37, 45, 46]. Based on this idea, a method for discovering subnetworks is as follows: start by selecting a threshold α of significant confidence, then construct a graph over variables, with an edge between x and y if this Markov pair is confident, i.e., $P(x \rightarrow y) > \alpha$. Each non-trivial component is considered as a seed of a subnetwork. Finally, expand the seed by adding variables that are related to this seed by a Markov pair with confidence above α', where $\alpha' < \alpha$ is an additional parameter.

The authors [13] proposed a new Bayesian score-based method. Assume a subnetwork contain k genes and l edges whose confidence $\{c_i\}_{i=1}^l$ is bigger than a threshold α. The score of a subnetwork is defined as $\prod_{i=1}^l c_i$. To find high scoring subnetworks, a greedy hill-climbing search was employed. This search starts with some seeds, say some nets three genes with two high scoring edges. Then add or remove a gene to the current subnetwork at each step. Once it reaches a local minimal, and then repeats the procedure until all seeds are used. Finally the subnetworks with highest scores are obtained.

2.4. *Inferring transcriptional regulatory networks based on integrated expression and sequence data*

Reference 45 presents an approach to understanding transcriptional regulatory based on integrated expression and sequence data. Here we briefly overview this approach. Assume the genes are partitioned into a set of K mutually exclusive modules using clustering. Hence, each gene is associated with an attribute $M \in \{1,....,K\}$ whose value represents the module to which the gene belongs. For each gene x, denote expression measurements as $x.E_1, x.E_2,..., x.E_J$, where $x.E_j$ represents the expression level in condition j. Assume the expression measurements are conditionally independent given the module assignment, say

$$P(E_1,.E_2,.., E_J \mid M) = \prod_{j=1}^{J} P(E_j \mid M)$$

where $P(E_j \mid M = m)$ can be modeled using Gaussian probability density function. Since pure clustering has difficulties to capture the regulatory relationship among genes, motifs in sequence data are employed to refine the module assignment [45].

Usually binding sites can be denoted by using position specific scoring matrix (PSSM). Suppose we are searching for motifs of length k, a PSSM w is a $p \times 4$ matrix, assigning for each position $1, 2,..., p$ (p-mer) and letter $l \in \{A,C,G,T\}$ a weight $w_i[l]$. For a gene x, we have a set of binary variable $x.R_1, x.R_2,..., x.R_L$ where $x.R_i \in \{true, false\}$ or $\{1,0\}$ denotes whether the motif TF regulates the gene or not. Furthermore, each gene x has a promoter sequence $x.S_1, x.S_2,..., x.S_n$, where $S_i \in \{A,C,G,T\}$. The probability of binding given the sequence is then specified by

$$P(x.S_1, x.S_2,..., x.S_n) = \text{logit}\left(\log\left(\frac{w_0}{n-p+1} \sum_{i=1}^{n-p+1} \exp\left\{ \sum_{i=1}^{p} w_i[S_{i+j-1}] \right\} \right) \right) \quad (8)$$

where w_0 is a threshold [45]. Actually only part of TFs play a regulation in each module, so a motif profile of a transcriptional module is employed to address the problem. A motif profile of one module m is defined as a set of weights u_{mi}, one for each motif, such that u_{mi} specifies the extent to which motif i plays a regulatory role in module m. Gene x with a module m is modeled by

$$P(x.M = m \mid R_1 = r_1,...,R_L = r_L) = \frac{\exp\left\{\sum_{i=1}^{L} u_{mi} r_i\right\}}{\sum_{k=1}^{K} \exp\left\{\sum_{i=1}^{L} u_{ki} r_i\right\}} \qquad (9)$$

where $\{r_i\}_{i=1}^{L} \in \{0,1\}$. The above three components are put together as a probabilistic graphical model. The model defines the following joint distribution [45]:

$$P(x.R, x.M, x.E \mid x.S) = \prod_{i=1}^{L} P(x.R_i \mid x.S) P(x.M \mid x.R) \prod_{j=1}^{J} P(x.E_j \mid x.M),$$

$$(10)$$

where $E = \{E_1,....,E_J\}$ and $S = \{S_1,....,S_n\}$. Maximizing the likelihood, we can refine the parameters M and R for each gene x. The expectation maximization (EM) algorithm can be used to estimate the model parameters [45].

According to the above four approaches, the key problem is how to determine the influence from one gene to another gene. It is a gene prediction and selection problem. Next we will present two approaches: one is a gene prediction approach based on multinomial probit regression with Bayesian gene selection, and the other is network construction based on clustering and MCMC predictor design.

3. Multinomial Probit Regression with Baysian Gene Selection

3.1. *Problem formulation*

Assume there are $n+1$ genes, say $x_1,...,x_n,x_{n+1}$. Without loss of generality, we assume the target gene is x_{n+1}, and let w denote this target gene. Then $\mathbf{w} = [w_1, w_2,..., w_m]^T$ denotes the normalized expression profiles of the target gene (for example, for the normalized ternary expression data, $w_j = 1$ indicates the sample j is up-regulated; $w_j = -1$

indicates the sample j is down-regulated; and $w_j = 0$ indicates the sample j is invariant). Denote

$$X = \begin{bmatrix} \text{Gene 1} & \text{Gene 2} & \cdots & \text{Gene 2} \\ x_{11} & x_{12} & \cdots & x_{1n} \\ x_{21} & x_{22} & \cdots & x_{2n} \\ \vdots & \vdots & \ddots & \vdots \\ x_{m1} & x_{m2} & \cdots & x_{mn} \end{bmatrix} \quad (11)$$

as the normalized expression profiles of genes $x_1,...,x_n$. The gene selection problem is to find some genes from $x_1,...,x_n$ that are useful in predicting some target gene w. Here we consider a more general case of gene prediction, i.e., assume that the gene expression profiles are normalized to K levels.

The perceptron has been proved to be an effective example to model the relationship between the target gene and the other genes [32]. Here we study this problem by using probit regression with Bayesian gene selection. Let X_i denote the ith row of matrix $\mathbf{X}$ in (11). In the binomial probit regression, i.e., when $K=2$, the relationship between w_i and the gene expression levels X_i is modeled a probit regressor [3] which yields

$$P(w_i = 1 | X_i) = \Phi(X_i \boldsymbol{\beta}), \quad i = 1,...,m \quad (12)$$

where $\boldsymbol{\beta} = (\beta_1, \beta_2, ..., \beta_n)^T$ is the vector of regression parameters and Φ is the standard normal cumulative distribution function. Introduce m independent latent variable $z_1, z_2, ..., z_m$, where $z_i \sim \mathbb{N}(X_i \boldsymbol{\beta}, 1)$, i.e.,

$$z_i = X_i \boldsymbol{\beta} + \mathbf{e}_i, \quad i = 1,...,m$$

and $e_i \sim \mathbb{N}(0,1)$, where $\mathbb{N}$ denotes a normal distribution. Define $\boldsymbol{\gamma}$ as the $n \times 1$ indicator vector with the jth element γ_j such that $\gamma_j = 0$ if $\beta_j = 0$ (the variable is not selected) and $\gamma_j = 1$ if $\beta_j \neq 0$ (the variable is selected). The Bayesian variable selection is to estimate $\boldsymbol{\gamma}$ from the posteriori distribution $p(\boldsymbol{\gamma} | \mathbf{z})$.

However, when $K>2$, the situation is different from the binormial case because we have to construct K-1 regression equations similar to (13). Introduce K-1 latent variables $z_1, z_2, ..., z_{K-1}$ and K-1 regression equations such that $z_k = \mathbf{X}\beta_k + e_k$, $k = 1, ..., K-1$ where $e_k \sim \mathbb{N}(0,1)$. Let z_k take m values $z_{k,1}, z_{k,2}, ..., z_{k,m}$. Denote $\mathbf{z}_k = [z_{k,1}, ..., z_{k,m}]^T$ and $\mathbf{e}_k = [e_{k,1}, ..., e_{k,m}]^T$. We then have

$$\mathbf{z}_k = \mathbf{X}\beta_k + \mathbf{e}_k \qquad (14)$$

This model is called the multinomial probit model. For background on multinomial probit models, see Ref. 29. Note that we do not have the observations of $\{\mathbf{z}_k\}_{k=1}^{K-1}$, which makes it difficult to estimate the parameters in (14). Here we discuss how to select the same strongest genes for the different regression equations. The model is a little different from (14), i.e., the selected genes do not change with the different regression equations. Note that the parameter β is still dependent on k and γ, denoted by $\beta_{k,\gamma}$. Then (14) is rewritten as

$$\mathbf{z}_k = \mathbf{X}_\gamma \beta_{k,\gamma} + \mathbf{e}_k, \quad k = 1, ..., K-1 \qquad (15)$$

where $\mathbf{X}_\gamma$ means the column of $\mathbf{X}$ corresponding to those elements of γ that are equal to 1, and similar comment to $\beta_{k,\gamma}$. Now the problem is how to estimate γ and the corresponding $\beta_{k,\gamma}$ and $\mathbf{z}_k$ for each equation in (15).

3.2. *Bayesian variable selection*

A Gibbs sampler is employed to estimate all the parameters. Given γ for equation k, the prior distribution of β_γ is $\beta_{k,\gamma} \sim \mathbb{N}(0, c(X_\gamma^T X_\gamma)^{-1})$, where c is a constant (we set $c=10$ in [63]). The detailed derivation of the posterior distributions of the parameters is given in Ref. 33. Here we summarize the procedure for Bayesian variable selection. Denote

$$S(\gamma, \mathbf{z}_k) = \mathbf{z}_k^T \mathbf{z}_k - \frac{c}{c+1} \mathbf{z}_k^T \mathbf{X}_\gamma (\mathbf{X}_\gamma^T \mathbf{X}_\gamma)^{-1} \mathbf{X}_\gamma^T \mathbf{z}_k, \quad k = 1, ..., K-1 \qquad (16)$$

Then the Gibbs sampling algorithm for estimation $\{\gamma, \beta_k, \mathbf{z}_k\}$ is as follows. By straightforward computing, the posteriori distribution $p(\gamma \mid \mathbf{z}_1, ..., \mathbf{z}_{K-1})$ is approximated by:

$$
\begin{aligned}
p(\gamma \mid \mathbf{z}_1, ..., \mathbf{z}_{K-1}) &\propto p(\mathbf{z}_1, ..., \mathbf{z}_{K-1} \mid \gamma) P(\gamma) \\
&\propto (1+c)^{-\frac{(K-1)n_\gamma}{2}} \exp\left\{-\frac{1}{2}\sum_{k=1}^{K-1} S(\gamma, \mathbf{z}_k)\right\} \prod_{i=1}^{n} \pi_i^{r_i} (1-\pi_i)^{1-r_i}
\end{aligned}
$$

(17)

and the posterior distribution $p(\beta_{k,\gamma} \mid \mathbf{z}_k)$ is $p(\beta_{k,\gamma} \mid \mathbf{z}_k)$ given by $\beta_{k,\gamma} \mid \mathbf{z}_k, \mathbf{X}_\gamma \sim N(V_\gamma \mathbf{X}_\gamma^T \mathbf{z}_k, V_\gamma)$. The Gibbs sampling algorithm for estimating $\gamma, \{\beta_{k,\gamma}\}, \{\mathbf{z}_k\}$ is as follows:

(1) Draw γ from $p(\gamma \mid \mathbf{z}_1, ..., \mathbf{z}_{K-1})$. We usually sample each γ_j independently from

$$
\begin{aligned}
p(\gamma_i \mid \mathbf{z}_1, ..., \mathbf{z}_{K-1}) &\propto p(\mathbf{z}_1, ..., \mathbf{z}_{K-1} \mid \gamma) P(\gamma_i) \\
&\propto (1+c)^{-\frac{(K-1)n_\gamma}{2}} \exp\left\{-\frac{1}{2}\sum_{k=1}^{K-1} S(\gamma, \mathbf{z}_k)\right\} \pi_i^{r_i} (1-\pi_i)^{1-r_i}
\end{aligned}
$$

(18)

where $n_\gamma = \Sigma_{j=1}^{n}\gamma_j$ and $c=10$ in this study, where $\pi_j = P(\gamma_j = 1)$ is a prior probability to select the jth gene. It is set as $\pi_j = 8/n$ according to the very small sample size. If π_j takes a larger value, then we found that oftentimes $(\mathbf{X}_\gamma^T \mathbf{X}_\gamma)^{-1}$ does not exist.

(2) Draw β_k from

$$
p(\beta_k \mid \gamma, \mathbf{z}_k) \propto N(V_\gamma \mathbf{X}_\gamma^T \mathbf{z}_k, V_\gamma), \quad \text{where } V_\gamma = \frac{c}{1+c}(\mathbf{X}_\gamma^T \mathbf{X}_\gamma)^{-1}.
$$

(19)

(3) Draw $\mathbf{z}_k = [\mathbf{z}_{k,1}, ..., \mathbf{z}_{k,m}]^T$, from a truncated normal distribution according to the following procedure [42]:

For $i = 1, 2, ..., m$

- if $w_i = k$, then draw $z_{k,i}$ according to $z_{k,i} \sim \mathbb{N}(\mathbf{X}_\gamma \boldsymbol{\beta}_k, 1)$ truncated left by $\max\limits_{j \neq k} z_{j,i}$, i.e.,

$$z_{k,i} \sim \mathbb{N}(\mathbf{X}_\gamma \boldsymbol{\beta}_k, 1) 1_{z_{k,i} > \max_{j \neq k} z_{j,i}} \tag{20}$$

- else $w_i = j$ and $j \neq k$, then draw $z_{j,i}$ according to $z_{j,i} \sim \mathbb{N}(\mathbf{X}_\gamma \boldsymbol{\beta}_j, 1)$ truncated right by the newly generated $z_{k,i}$, i.e.,

$$z_{j,i} \sim \mathbb{N}(\mathbf{X}_\gamma \boldsymbol{\beta}_j, 1) 1_{\{z_{j,i} \leq z_{k,i}\}} \tag{21}$$

endfor

Here we set $z_{K,i} \sim \mathbb{N}(0,1)$ when $w_i = k$, i.e., we introduce a new equation $z_{K,i} = \mathbf{X}_\gamma \boldsymbol{\beta}_K + e_{K,i}, i = 1, ..., m$ with $\boldsymbol{\beta}_K$ being a zero vector and $e_{K,i} \sim \mathbb{N}(0,1)$.

In this study, 12,000 Gibbs iterations are implemented with the first 2,000 as the burn-in period. Then we obtain the Monte Carlo samples as $\{\boldsymbol{\gamma}^{(t)}, \boldsymbol{\beta}_k^{(t)}, \mathbf{z}_k^{(t)}, t = 2001, ..., T\}$, where T=10000. Finally we count the number of times that each gene appears in $\{\boldsymbol{\gamma}^{(t)}, t = 2001, ..., T\}$. The genes with the highest appearance frequencies play the strongest role in predicting the target gene. We will discuss some implementation issues of this algorithm in the next section.

3.3. *Bayesian estimation using the strongest genes*

Now assume that the genes corresponding non-zeros of $\boldsymbol{\gamma}$ are the strongest genes obtained by the above Bayesian variable selection algorithm. For fixed $\boldsymbol{\gamma}$, we again use a Gibbs sampler to estimate the probit regression coefficients $\boldsymbol{\beta}_k$ as follows: First draw $\boldsymbol{\beta}_{k,\gamma}$ according to (19), then draw $\mathbf{z}_k$, and iterate the two steps. In this study, 1,500 iterations are implemented with the first 500 as the burn-in period. Thus we obtain the Monte Carlo samples $\{\boldsymbol{\beta}_{k,\gamma}^{(t)}, \mathbf{z}_k^{(t)}, t = 501, ..., \tilde{T}\}$. The probability of a given sample $\mathbf{x}$ under each class is given by:

$$P(w = k \mid \mathbf{x}) = \frac{1}{\tilde{T}} \sum_{t=1}^{\tilde{T}} \prod_{j=1,\, j \neq k}^{K} \Phi(\mathbf{x_r}\boldsymbol{\beta}_{k,\mathbf{r}}^{(t)} - \mathbf{x_r}\boldsymbol{\beta}_{j,\mathbf{r}}^{(t)}), \quad k = 1, \dots, K-1 \quad (22)$$

$$P(w = K \mid \mathbf{x}) = 1 - \sum_{k=1}^{K-1} P(w = k \mid \mathbf{x}) \quad (23)$$

where $\boldsymbol{\beta}_{K,\mathbf{r}}^{(t)}$ is a zero vector; and the estimation of this sample is given by:

$$\hat{w} = d(w) = \underset{1 \leq k \leq K}{\arg\max}\, P(w = k \mid \mathbf{x}) \quad (24)$$

Note that (22) may be computed using another formulation, see [58].

To measure the fitting accuracy of such a predictor, we next define the coefficients of determination (CoD) [59] for this probit predictor. In fact, the above γ and β (including all parameters $\boldsymbol{\beta}_{k,\gamma}$) are dependent on the target gene w. Firstly, a probabilistic error measure $\varepsilon(w, \mathbf{x}_\gamma, \boldsymbol{\beta})$ associated with the predictor γ, β is defined as $\varepsilon(w, \mathbf{x}_\gamma, \boldsymbol{\beta}) \triangleq \mathrm{E}\left[\mid d(w) - w \mid^2\right]$, where E denotes the expectation. Similar to the definition in [14], the CoD for w relative to the conditioning set γ, β is defined by:

$$\theta = \frac{\varepsilon - \varepsilon(w, \mathbf{x}_\gamma, \boldsymbol{\beta})}{\varepsilon}$$

where ε is the error of the best (constant) estimate of w in the absence of any conditional variables. In the case of minimum mean-square error estimation, ε is defined as $\varepsilon = \mathrm{E}\left[\mid w - g(\mathrm{E}(w) \mid^2\right]$, where g is a {-1,0,1}–valued threshold function [$g(z) = 0$ if $-0.5 < z < 0.5$, $g(z) = 1$ if $z \geq 0.5$, and $g(z) = -1$ if $z \leq -0.5$] for ternary data.

3.4. *Experimental results*

As the first step in constructing a gene regulatory network, the complexity of the expression data is reduced by thresholding changes in the transcription level into ternary expression data: -1 (down-regulated), +1 (up-regulated), or 0 (invariant). When using multiple microarrays, the absolute signal intensities vary extensively due to both the process of preparing and printing the EST elements [11], and the process of preparing and labeling the cDNA representations of the RNA pools. This problem is solved via internal standardization. We then build gene regulatory networks using the proposed approaches.

The gene-expression profiles used in this study result from a study of 31 malignant melanoma samples [8]. For the study, total messenger RNA was isolated directly from melanoma biopsies. Fluorescent cDNA from the message was prepared and hybridized to a microarray containing probes for 8,150 cDNAs (representing 6, 971 unique genes). A set of 587 genes has been subjected to an analysis of their ability to cross-predict each other's state in a multivariate setting [32]. From these, we have selected twenty-six differential genes using t-test [56]. The genes are listed in Table 1.

CoD values for all 26 targets have been computed using the strongest genes found via the Bayesian selection. CoDs have been computed using the leave-one-out cross validation. The strongest genes for each target are listed in the second column of Table 2. The third column lists the CoDs also use the top 2, 3, and 4 genes for each target and the probit regression to form the predictors. Several points should be noted. First, while the theoretical (distributional) CoD values will increase as the number of predictors increase, this is not necessarily the case for experimental data, especially when small samples are involved (on account of over-fitting and high variance of cross-validation error estimation). Second, pirin (no. 2) is a strong predictor gene in many cases, and this agrees with the comment in the original paper that pirin has a high discriminative weight [8]. Third, even with feature selection and a suboptimal predictor function, for the most part the CoDs are fairly high.

Based on the above gene prediction and selection procedure, it is easy to construct regulatory subnetworks. The procedure is similar to that in Section 2.3. In the meantime, we can also construct networks using the influence defined by the appearing frequencies of each gene to predict target gene. This is our future work.

Table 1. The 26 differential genes.

Gene No	Index No	Gene description
1	3	tumor protein D52
2	7	pirin
3	14	v-myc avian myelocytomatosis viral oncogene homolog
4	42	endothelin receptor type B
5	60	ESTs
6	79	alpha-2-macroglobulin
7	117	v-myc avian myelocytomatosis viral oncogene homolog
8	126	ESTs
9	175	myotubularin related protein 4
10	210	NGFI-A binding protein 2 (ERG1 binding protein 2)
11	216	IQ motif containing GTPase activating protein 1
12	220	annexin A2
13	228	annexin A2
14	245	Homo sapiens mRNA; cDNA DKFZp434L057 (from clone KFZp434L057)
15	282	endothelin receptor type B
16	292	ESTs
17	323	ESTs
18	360	glycoprotein M6B
19	372	nuclear receptor subfamily 4, group A, member 3
20	374	thrombospondin 2
21	387	ESTs, Weakly similar to HP1-BP74 protein [M.musculus]
22	404	phosphofructokinase, liver
23	506	placental transmembrane protein
24	556	Human insulin-like growth factor binding protein 5 (IGFBP5) mRNA
25	573	platelet-derived growth factor receptor, alpha polypeptide
26	576	ESTs

Table 2. The strongest genes to predict each gene and the corresponding CoD values for 2, 3, and 4 predictors.

Target genes (No.)	Strongest genes (No.)				COD		
	1	2	3	4	2	3	4
1	19	23	22	17	0.6452	0.6129	0.7097
2	25	1	19	11	0.3871	0.6774	0.8065
3	7	23	2	5	0.7097	0.7742	0.7742
4	15	2	13	17	0.7419	0.7742	0.8710
5	14	2	13	10	0.5484	0.5161	0.4194
6	10	2	19	24	0.6129	0.7097	0.8387
7	3	2	17	1	0.7419	0.8387	0.8387
8	20	2	21	14	0.5161	0.5484	0.5484
9	2	13	17	15	0.6774	0.7097	0.7742
10	6	20	2	4	0.6129	0.6452	0.6774
11	13	25	2	1	0.8710	0.8710	0.7742
12	2	13	11	14	0.6452	0.6452	0.7419
13	2	15	11	18	0.8387	1.0000	1.0000
14	2	25	21	15	0.6774	0.7742	0.6774
15	2	4	13	14	0.8065	0.7419	0.9677
16	4	25	2	7	0.6452	0.7097	0.6452
17	11	18	2	8	0.8387	0.8065	0.8387
18	2	17	13	23	0.8387	0.7742	0.8710
19	1	22	2	9	0.7419	0.6774	0.7419
20	22	5	10	24	0.3548	0.3548	0.7419
21	25	2	14	20	0.7742	0.7742	0.7742
22	2	9	6	23	0.6774	0.7097	0.7742
23	24	2	1	5	0.5161	0.5484	0.6774
24	2	20	3	7	0.5806	0.6129	0.6452
25	11	2	14	13	0.7742	0.6774	0.8065
26	17	13	2	23	0.7742	0.7742	0.838

4. Network Construction Based on Clustering and Predictor Design

Strictly speaking, the approach in Section 2 is a gene regulatory network inference. So here we also present our related network construction method. Clustering analysis [62] is an important approach in gene regulatory pathway study. In [60], mutual-information-based clustering is used as the first step in network construction. Then, the set of parent genes for each gene could be obtained by using the clustering procedure. Then, it is natural to define the $l(i)$ sets $\{X_k^{(i)}\}_{k=1}^{l(i)}$ as influencing sets for gene x_i to be composed of the other genes in clusters $C_1, C_2, ..., C_{l(i)}$. For the moment, let us proceed under this assumption. Later on we will make an adjustment on this assumption.

4.1. *Predictor construction using reversible jump MCMC annealing*

Based on the possible $l(i)$ sets $\{X_k^{(i)}\}_{k=1}^{l(i)}$, we want to estimate the predictors $f_k^{(i)} : X_k^{(i)} \to y_i, k = 1, ..., l(i)$ from the observations (in fact, y_i should be x_i, but to describe the functions clearly we let y_i stand for x_i). For notational convenience, we consider a general formulation, $f : X \to y$. We postulate a multivariate-input, single-output mapping $y_t = f(\mathbf{x}_t) + e_t$ where $\mathbf{x}_t \in \mathbb{N}^d$ is a set of input variables, $y_t \in \mathbb{N}$ is a target variable, e_t is a noise term, and $t \in \{1, 2, ...\}$ is an indexed variable over the data. Here we slightly abuse the notation, i.e., denote $\mathbb{N}$ as the integer set. The learning problem involves computing an approximation to the function f and estimating the characteristics of the noise process given a set of m input-output observations $O = \{\mathbf{x}_1, \mathbf{x}_2, ..., \mathbf{x}_m; y_1, y_2, ..., y_m\}$.

We consider the following family of predictors:

$$M_0 : \quad y_t = b + \boldsymbol{\beta}^T \mathbf{x}_t + e_t \tag{25}$$

$$M_1 : \quad y_t = \sum_{j=1}^{J} a_j \phi(\| \mathbf{x}_t - \boldsymbol{\mu}_t \|) b + \boldsymbol{\beta}^T \mathbf{x}_t + e_t \tag{26}$$

where $1 \le J \le J_{\max}$ and $J_{\max}$ is an upper-bound on the number of nonlinear terms in (26); ϕ is a radial basis function (RBF); $\| \cdot \|$ denotes a distance metric (usually Euclidean or Mahalanobis); $\boldsymbol{\mu}_j \in \mathbb{N}^d$ denotes the jth RBF center; $a_j \in \mathbb{N}$ is the jth RBF coefficient; $b \in \mathbb{N}$; $\boldsymbol{\beta} \in \mathbb{N}^d$ are the linear regression parameters; and $e_t \in \mathbb{N}$ is assumed to be i.i.d. noise. Depending on our *a priori* knowledge about the smoothness of the mapping, we can choose different types of basis functions. Here the Gaussian basis function is used, given by $\phi(\rho) = \exp(-\rho^2)$. Denote $\mathbf{y} = [y_1, ..., y_m]^T$, $\alpha = [b, \beta_1, ..., \beta_d, a_1, ..., a_J]^T$, $\mathbf{e} = [e_1, ..., e_m]^T$ and

$$D \triangleq \begin{bmatrix} 1 & x_{1,1} & \cdots & x_{1,d} & \phi(\| \mathbf{x}_1 - \boldsymbol{\mu}_1 \|) & \cdots & \phi(\| \mathbf{x}_1 - \boldsymbol{\mu}_J \|) \\ 1 & x_{2,1} & \cdots & x_{2,d} & \phi(\| \mathbf{x}_2 - \boldsymbol{\mu}_1 \|) & \cdots & \phi(\| \mathbf{x}_2 - \boldsymbol{\mu}_J \|) \\ \vdots & \vdots & \ddots & \vdots & \vdots & & \ddots & \vdots \\ 1 & x_{m,1} & \cdots & x_{m,d} & \phi(\| \mathbf{x}_m - \boldsymbol{\mu}_1 \|) & \cdots & \phi(\| \mathbf{x}_m - \boldsymbol{\mu}_J \|) \end{bmatrix} \quad (27)$$

We express the model (26) in vector-matrix form as:

$$\mathbf{y} = D\alpha + \mathbf{e} \qquad (28)$$

The noise variance is assumed to be σ^2. We assume here that the number J of RBFs and the parameters $\Theta_J \triangleq \{\boldsymbol{\mu}_1, ..., \boldsymbol{\mu}_J, \sigma^2, \alpha\}$ are unknown. Given the data set O, our objective is to estimate J and Θ_J, where $\Theta_J \in \mathbb{R}^\kappa \times \mathbb{R}^+ \times \Omega_J$ with $\kappa = 1 + d + J$; that is $\alpha \in \mathbb{R}^\kappa$, $\sigma^2 \in \mathbb{R}^+$, and $\boldsymbol{\mu} \in \Omega_J$, where Ω_J is defined as $\Omega_J = \left\{ \boldsymbol{\mu} : \mu_{j,i} \in \left[\min_{1 \le l \le m} x_{l,i} - 0.5, \max_{1 \le l \le m} x_{l,i} - 0.5 \right] \right\}$.

The Bayesian inference of J and Θ_J is based on the joint posterior distribution $p(J, \Theta_J | O)$. Our aim is to estimate the joint distribution from which, by standard probability marginalization and transformation techniques, one theoretically expresses all posterior features of interest. It is not possible to obtain these quantities analytically, since this would require the evaluation of high dimensional integrals of nonlinear functions in the parameters. Consequently, we propose to use a Markov chain Monte Carlo (MCMC) method to perform Bayesian computation. The basic idea of MCMC is to build an ergodic Markov chain $\{J^{(i)}, \Theta_J^{(i)}, i \in \mathbb{N}\}$ whose equilibrium distribution is the desired joint posterior distribution $p(J, \Theta_J | O)$.

From model (28), given $J, \mu_1, ..., \mu_J$, the least-squares estimate of α is given by

$$\hat{\alpha} = [\mathbf{D}^T \mathbf{D}]^{-1} \mathbf{D}^T \mathbf{y} \tag{29}$$

An estimate of σ^2 is then given by

$$\hat{\sigma}^2 = \frac{1}{m} (y - \mathbf{D}\hat{\alpha})^T (y - \mathbf{D}\hat{\alpha}) = \frac{1}{m} y \mathbf{P}_J^* y \tag{30}$$

where $\mathbf{P}_J^* = \mathbf{I}_m - \mathbf{D}[\mathbf{D}^T \mathbf{D}]^{-1} \mathbf{D}^T$. Base on the minimum description length (MDL) criterion, we can impose the following *a priori* distribution on J [4],

$$P(J) \propto \exp\left[-\left(J + \frac{d+1}{2} \right) \log m \right] \tag{31}$$

Assuming the noise samples are i.i.d. Gaussian, it can then be shown that the joint posterior distribution of $(J, \mu_1, ..., \mu_J)$ is given by [4],

$$p(J, \mu_1, ..., \mu_J) \propto \left[(y^T \mathbf{P}_J^* y)^{-\frac{m}{2}} \right] P(J) \tag{32}$$

Hence the maximum *a posteriori* (MAP) estimate of these parameters is obtained by maximizing the right-hand side of (32). This can be done by using the reversible jump MCMC algorithm [44].

4.2. *CoD for predictors*

A natural way to select a set of predictors for a given gene is to employ the *coefficients of determination* (CoD) [59] Assume $X_1^{(i)}, X_2^{(i)}, ..., X_{l(i)}^{(i)}$ to be sets of genes; and $f_1^{(i)}, f_2^{(i)}, ..., f_{l(i)}^{(i)}$ to be function rules that are estimated by the reversible jump MCMC method developed in Section 4.2. Then the probabilistic error measure $\varepsilon(x_i, f_k^{(i)}(X_k^{(i)}))$ is defined as:

$$\varepsilon(x_i, f_k^{(i)}(X_k^{(i)})) \triangleq \mathbf{E}\left[\left| g(f_k^{(i)}(X_k^{(i)})) - x_i \right|^2 \right] \tag{33}$$

where g is defined as in the last section. For each k, the CoD for x_i relative to the conditioning set $X_k^{(i)}$ is defined by [48]:

$$\theta_k^{(i)} = \frac{\varepsilon_i - \varepsilon(x_i, f_k^{(i)}(X_k^{(i)}))}{\varepsilon_i}, \quad k = 1, \ldots, l(i) \tag{34}$$

where ε_i is the error of the best (constant) estimate of x_i in the absence of any conditional variables, and here it is defined as $\varepsilon_i = E\left[\left|x_i - g(\mathbf{E}(x_i))\right|^2\right]$.

Once a class of gene sets $X_1^{(i)}, X_2^{(i)}, \ldots, X_{l(i)}^{(i)}$ has been selected based on clustering results, we take the designed approximations of the optimal function rules $f_1^{(i)}, f_2^{(i)}, \ldots, f_{l(i)}^{(i)}$ as the rule for gene x_i with the probability of $f_k^{(i)}$ being $c_k^{(i)} = \theta_k^{(i)} / \sum_{j=1}^{l(i)} \theta_j^{(i)}, k = 1, \ldots, l(i)$, where the CoDs are the estimates formed from the training data according to (34}). The number $l(i)$ of predictors is based on the optimization of the number of clusters. Suppose the clusters containing gene x_i are $C_1, C_2, \ldots, C_{l(i)}$. Owing to the difficulty of estimating entropy with small sample sizes, there is potential imprecision in these clusters. Consequently, when sample sizes are small, as they are in the application we will show next, we will proceed in a more flexible manner in choosing the $l(i)$ sets. Suppose $C_\kappa = \{x_i, g_{\kappa,1}, g_{\kappa,2}, \ldots, g_{\kappa,r(\kappa)}\}$, where $r(\kappa)$ is the number of genes in C_κ besides x_i. For $j = 1, 2, \ldots, r(\kappa)$, form the sets $\{C_k, C_{\kappa,j1}, C_{\kappa,j2}, \ldots, C_{\kappa,j,s(\kappa)}\}$, where $s(\kappa)$ is the number of genes not in C_κ, by interchanging $g_{\kappa,j}$ with each of the genes not in C_κ. We then choose $f_\kappa^{(i)}$ to be the predictor with highest CoD from among the predictors formed from $\{C_k, C_{\kappa,j1}, C_{\kappa,j2}, \ldots, C_{\kappa,j,s(\kappa)}\}, j = 1, 2, \ldots, r(\kappa)$.

4.3. *Experimental results on a Myeloid line*

The ability of the mutual-information clustering algorithm and MCMC predictors to build probabilistic gene regulatory networks based on the observations in transcription level has been tested in the context of responsiveness to genotoxic stresses. The cell lines were chosen so that a sampling of both *p53* proficient and *p53* deficient cells would be assayed. By using the same data set used for the initial studies concerning nonlinear prediction in the context of genomic microarray data [32], we are able to compare the results of the proposed, information-MCMC approach with the previously obtained results. The ternary data of the survey (14 genes and 30 samples) are given in [32], where the conditions *IR, MMS,* and *UV* have the values 1 or 0, depending on whether the condition is or is not in effect. The 14 genes are *RCH1* (x_1), *BCL3* (x_2), *FRA1* (x_3), *REL-B* (x_4), *ATF3* (x_5), *IAP-1* (x_6), *PC-1*(x_7), *MBP-1* (x_8), *SSAT* (x_9), *MDM2* (x_{10}), *p21* (x_{11}), *p53* (x_{12}), *AHA* (x_{13}), and *OHO* (x_{14}). To maintain consistency with [32], we continue to employ the blind controls used in those studies, in which expression patterns for two fictitious genes were created.

The optimal partition number is four, and we select the *k*-partition schemes for *k*=2 to *k*=7. From the clustering results in [59], we can find the possible parents for each gene. Note that the artificial gene *AHA* (x_{13}) has been defined so that it is tightly coupled to *p53* (x_{12}). For the details of the probabilistic gene regulatory networks based on this data set, we refer to [59]. Here we compare the results of the new information-MCMC approach with those obtained by using a full search restricted by a small number of predictor genes. To do so, we consider one target gene, *p53* (x_{12}), whose predictor sets $X_1^{(12)}$ and are shown in Table 3. Here we make some comparisons with results reported in [32]. In both studies, the CoD of a predictor is estimated using cross-validation among the 30 samples: 20 samples are randomly chosen for training, the remaining 10 samples are used for test data, and this is performed *1000* times to estimate the CoD. It is known that *p53* (x_{12}) is influential, but

Table 3. Regulatory rules for gene *p53* (x_{12}).

parent sets	Reversible jump MCMC		Perceptron	
	CoD	$c_{l(12)}^{(12)}$	CoD	$c_{l(12)}^{(12)}$
$X_1^{(12)} = \{x_{13}\}$	0.5532	0.4764	0.5532	0.5051
$X_1^{(12)} = \{x_5, x_{10}, x_{11} x_{13}\}$	0.6080	0.5236	0.5421	0.4949

not determinative, of the up regulation of both *p21* (x_{11}) and *MDM2*
(x_{10}). Thus, some level of prediction of *p53* should be possible by a
combination of these two genes. This expectation is met with *p21* and
MDM2 combining with *ATF3* (x_5) and *AHA* (x_{13}) as shown in Table 3.
To fully interpret the results, we must recognize that *AHA* is an artificial
gene defined to couple with *p53*. This is evidenced in 3, which shows
that, acting alone as $X_1^{(12)}$, *AHA* predicts *p53* with a substantial CoD.
Adjoining *p21*, *MDM2*, and *ATF3* increases the MCMC CoD, whereas
the perceptron CoD is unchanged (the small variation in empirical error
being insignificant). We conjecture that the increment is due to *p21* and
MDM2. One should not be surprised to see the inclusion of a
noninfluential gene, such as we conjecture *ATF3* to be, in predictor sets.
Not only is this inevitable when *k* is small, but it also reflects that not all
related genes are co-expressed and some unrelated genes have similar
expression patterns [26]. The key point is that the set {*p21*, *MDM2*,
ATF3, *OHO*} forms a good predictor set for the probabilistic gene
regulatory networks.

5. Concluding Remarks

In this chapter, we have briefly reviewed transcriptional regulatory
networks, inferring subnetworks using using Z-score, graph theory, and
Bayesian networks, as well as inferring transcriptional regulatory
networks based on integrated expression and sequence data. We studied
the problem of multi-level gene prediction and genetic network
construction from gene expression data based on multi-nomial probit
regression with Bayesian gene selection that selects genes closely related
to a particular target gene. Experimental results from malignant
melanoma samples have been shown that this approach can effectively
discover the relationship among genes. Based on the Bayesian gene
prediction and selection, we can construct subnetworks from microarray
data. We also discussed a probabilistic gene regulatory network
construction method using clustering and MCMC predictor design. We
have provided experimental results on microarray data from known gene
response pathways including ionizing radiation and downstream targets
of inactivating gene mutations.

Although there are many approaches to constructing gene regulatory networks, this area of research is just in its infancy. For example, recent findings indicate that not all co-influenced genes have the same transcriptional factors or co-regulated [6, 53]. Therefore, there is still a long way to go to solve the network construction problem.

References

[1] T. Akutsu, S. Miyano, and S. Kuhara, "Identification of genetic networks from a small number of gene expression patterns under Boolean network model," Pacific Symposium on Biocomputing, Vol. 4, pp. 17-28, 1999.

[2] T. Akutsu, S. Miyano, and S. Kuhara, "Inferring qualitative relations in genetic networks and metabolic pathways," Bioinformatics, Vol. 16, pp. 727-743, 2000.

[3] J. Albert and S. Chib, "Bayesian analysis of binary and polychotomous response data," \emph{Journal of the American Statistical Association}, Vol. 88, pp. 669-679, 1993.

[4] C. Andrieu, J. Freitas, and A. Doucet, "Robust full Bayesian learning for neural networks," Neural Computation, Vol. 13, pp. 2359-2407, 2001.

[5] M.N. Arbeitman, E.E.M. Furlong, F. Imam, E. Johnson, B.H. Null, B.S. Baker, M.A. Krasnow, M.P. Scott, R.W. Davis, and K.P. White, "Gene expression during the life cycle of Drosophila melanogaster," Science, Vol. 297, pp. 2270-2275, 2002.

[6] S. Li, C.M. Armstrong, *et al.*, "A map of the interactome network of the metazoan C. elegans," Science, Vol. 303, pp. 540-543, 2004.

[7] A. Bhan, D.J. Galas, and T.G. Dewey, "A duplication frowth model of gene expression networks," Bioinformatics, Vol. 18, pp. 1486-1493, 2002.

[8] M. Bittner, P. Meltzer, *et al.*, "Molecular classification of cutaneous malignant melanoma by gene expression profiling," Nature, Vol. 406, pp.536-540, 2000.

[9] H.J. Bussemaker, H. Li, and E.D. Siggia, "Regulatory element detection using correlation with expression," Nature Genetics, Vol. 27, pp. 167-171, 2001.

[10] T. Chen, V. Filkov, and S. Skiena, "Identifying gene regulatory networks from experimental data," In Proc. RECOMB, pp. 94-103, 1999.

[11] Y. Chen, E.R. Dougherty, and M.L. Bittner, "Ratio-based decisions and the quantitive analysis of cDNA microarray images," Journal of Biomedical Optics, Vol. 2, No. 4, pp. 364-374, 1997.

[12] Y. Chen, V. Kamat, E.R. Dougherty, M. Bittner, P.S. Meltzer, and J.M. Trent, "Ratio statistics of gene expression levels and applications to mocroarray data analysis," Bioinformatics, Vol. 18, pp. 1207-1215, 2002.

[13] D. Pe'er, A. Regev, G. Elidan, and N. Friedman, "Inferring subnetworks from perturbed expression profiles," Bioinformatics, Vol. 17, pp. 215S-224S, 2001.

[14] E.R. Dougherty, S. Kim, and Y. Chen, "Coefficient of determination in nonlinear signal processing," Signal Processing, Vol. 80, pp. 2219-2235, 2000.

[15] N. Friedman, M. Linial, I. Nachman, and D. Pe'er, "Using Bayesian network to analyze expression data," Journal of Computational Biology, Vol. 7, pp. 601-620, 2000.

[16] N. Friedman and D. Koller, "Being Bayesian about network structure: A Bayesian approach to structure discovery in Bayesian networks," Machine Learning, Vol. 50, pp. 95-126, 2003.

[17] D. Geiger and D. Heckerman, "Parameter priors for directed acyclic graphical models and the characterization of several probability distributions," Annals of Statistics, Vol.31, pp. 1412-1440, 2002.

[18] D. Giudici and R. Castelo, "Improving Markov chain Monte Carlo model search for data mining," Machine Learning, Vol. 50, pp. 127-158, 2003.

[19] A.J. Glass nad S.A. Kauffman, "The logical analysis of continuous, non-linear biochemical networks," Journal of Theoretical Biology, Vol. 39, 103-129, 1973.

[20] T.R. Golub, D.K. Slonim, P. Tamayo, C. Huard, M. Gaasenbeek, J.P. Mesirov, H. Coller, M.L. Loh, J.R. Downing, M.A. Caligiuri, C.D. Bloomfield, and E.S. Lander, "Molecular classification of cancer: class discovery and class prediction by gene expression monitoring," Science, Vol. 286, pp. 531-537, 1999.

[21] P. D'Haeseleer, X. Wen, S. Fuhrman, and R. Somogyi, "Linear modeling of mRNA expression levels during CNS development and injury," Pacific Symposium on Biocomputing, Vol. 4, pp. 41-52, 1999.

[22] P. D'Haeseleer, P. Liang, S. Fuhrman, and R. Somogyi, "Genetic network inference: from co-expression clustering to reverse engineering," Bioinformatics, Vol. 16, pp. 707-726, 2000.

[23] A.J. Hartemink, D.K. Gifford, T.S. Jaakkola, and R.A. Young, "Using graphical models and genomic expression data to statistically validate models of genetic regulatory networks," Pacific Symposium on Biocomputing, Vol. 6, pp. 23-32, 2001.

[24] J. Hasty, D. McMillen, F. Isaacs, and J.J. Collins, "Computational studies of gene regulatory networks: in numero molecular biology," Nature Reviewers Genetics, Vol. 2, pp. 268-279, 2001.

[25] D. Heckerman, "A tutorial on learning with Bayesian networks," http://research.microsoft.com/~heckerman, 1996.

[26] L.J. Heyer, S. Kruglyak, and S. Yooseph, "Exploring expression data: identification and analysis of coexpressed genes," Genome Research, Vol. 9, pp. 1106-1115, Aug. 1999.

[27] S. Huang, "Gene expression profiling, genetic networks, and cellular states: an integrating concept for tumorigenesis drug biology," Journal of Molecular Medicine, Vol. 77, pp. 469-480, 1999.

[28] T. Ideker, O. Ozier, B. Schwirkowski, and A.F. Siegel, "Discovering regulatory and signaling circuits in molecular interaction networks," \emph{Bioinformatics, Vol. 18, pp. S233-S240, 2002.

[29] K. Imai and D.A. van Dyk, "A Bayesian analysis of the multinomial probit model using marginal data augmentation," http://www.people.fas.harvard.edu/~kimai/research/mnp.html

[30] S.A. Kauffman, "Metabolic stability and epigenesis in randomly constructed genetic nets," Journal of Theoretical Biology, Vol. 22, 437-467, 1969.

[31] S.A. Kauffman, The origins of order: Self-organization and selection in evolution, Oxford University Press, New York, 1993.

[32] S. Kim, E.R. Dougherty, M.L. Bittner, Y. Chen, K. Sivakumar, P. Meltzer, and J.M. Trent, "General nonlinear framework for the analysis of gene interaction via multivariate expression arrays," Journal of Biomedical Optics, Vol. 5, pp. 411-424, 2000.

[33] K.E. Lee, N. Sha, E.R. Dougherty, M. Vannucci, and B.K. Mallick, "Gene selection: a Bayesian variable selection approach," Bioinformatics, Vol. 19, pp. 90-97, 2003.

[34] T. Lee, N. Rinaldi, F. Robert, D. Odom, Z. Bar-Joseph, G. Gerber, N. Hannett, C. Harbison, C. Thompson, I. Simon, J. Zeitlinger, E. Jennings, H. Murray, D. Gordon, B. Ren, J. Wyrick, J. Tagne, T. Volkert, E. Fraenkel, D. Gifford, R. Young, "Transcriptional regulatory networks in Saccharomyces cerevisiae," Science, Vol. 298, pp. 799-804. 2002.

[35] T. Mestl, E. Plahte, and S.W. Omholt, "A mathematical framework for describing and analysing gene regulatory networks," Journal of Theoretical Biology, Vol. 176, pp. 291-300, 1995.

[36] R. Milo, S. Shen-Orr, S. Itzkovitz, N. Kashtan, D. Chklovskii, U. Alon, "Network motifs: simple building blocks of complex networks," Science, Vol. 298, pp. 824-827, 2002.

[37] E.J. Moler, D.C. Radisky, and I.S. Mian, "Integrating naive Bayes models and external knowledge to examine copper and iron homeostasis in S. cerevisiae," Physiologyical Genomics, Vol. 4, pp. 127-135, 2000.

[38] Y. Pilpel, P. Sudarsanam, and G.M. Church, "Identifying regulatory networks by combinatorial analysis of promoter elements," Nature Genetics, Vol. 29, pp. 153-159, 2001.

[39] J. Qian, J. Lin, N.M. Luscombe, H. Yu, and M. Gerstein, "Prediction of regulatory networks: genome-wide identification of transcription factor targets from gene expression data," Bioinformatics, Vol. 19, pp. 1917-1926, 2003.

[40] Z.S. Qin, L.A. McCue, W. Thompson, L. Mayerhofer, C.E. Lawrence, and J.S. Liu, "Identification of co-regulated genes through Bayesian clustering of predicted regulatory binding sites," Nature Biotechnology, Vol. 21, pp. 435-439, 2003.

[41] J. Rissanen, "Stochastic complexity," Journal of the Royal Statistical Society, Vol. 49, pp. 223-239. 1987.

[42] C. Robert, "Simulation of truncated normal variables," Statistics and Computing, vol. 5, pp. 121-125, 1995.

[43] C.P. Robert amd G. Casella, Monte Carlo statistical methods, Springer-Verlag Press, New York, 1999.

[44] G.A.F. Seber, Multivariate Observations, New York:Wiley, 1984.

[45] E. Segal, R. Yelensky, and D. Koller, "Genome-wide discovery of transcriptional modules from DNA sequence and gene expression," Bioinformatics, Vol. 19, pp. S273-282, 2003.

[46] E. Segal, M. Shapira, A. Regev, D. Pe'er, D. Botstein, D. Koller, and N. Friedman, "Module networks: discovering regulatory modules and their condition specific regulators from gene expression data," Nature Genetics, Vol. 34, pp. 166-176, 2003.

[47] S.S. Shen-Orr, R. Milo, S. Mangan1, and U. Alon, "Network motifs in the transcriptional regulation network of Escherichia coli," Nature Genetics, Vol. 31, pp. 64-68, 2002.

[48] I. Shmulevich, E.R. Dougherty, S. Kim, and W. Zhang, "Probabilistic Boolean networks: a rule-based uncertainty model for gene regulatory networks," Bioinformatics, Vol. 18, pp. 261-274, 2002.

[49] P. Smolen, D. Baxter, and J. Byrne, "Mathematical modeling of gene networks," Neuron, Vol. 26, pp. 567-580, 2000.

[50] E.P. van Someren, L.F.A. Wessels, and M.J.T. Reinders, "Linear modeling of genetic networks from experimental data," Intelligent Systems for Molecular Biology, San Diego, 2000.

[51] Y. Tamada, S.Y. Kim, H. Bannai, S. Imoto, K. Tashiro, S. Kuhara, and S. Miyano, "Estimating gene networks from gene expression data by combining Bayesian network model with promoter element detection," Bioinformatics, Vol. 19, pp. 227ii-236ii, 2003.

[52] S. Tavazoie, J.D. Hughes, M.J. Campbell, R.J. Cho, and G.M. Church, "Systematic determination of genetic network architecture," Nature Genetics, Vol. 22, pp. 281-285, 1999.

[53] A.H.Y. Tong, G. Lesage, ..., F.P. Roth, G.W. Brown, B. Andrews, H. Bussey, and C. Boone, "Global mapping of the yeast genetic interaction network," Science, Vol. 303, pp. 808-813, 2004.

[54] O. Troyanskaya, M. Cantor, G. Sherlock, P. Brown, T. Hastie, R. Tibshirani, D. Botstein, and R.B. Altman, "Missing value estimation methods for DNA microarrays," Bioinformatics, Vol. 17, pp. 520-525, 2001.

[55] G.C. Tseng, M.K. Lars Rohlin, J.C. Liao, and W. H. Wong, "Issues in cDNA microarray analysis: quality filtering, channel normalization, models of variation and assessment of gene effects," Nucleic Acids Research, Vol. 29, pp. 2549-2557, 2001.

[56] R.E. Walpole and R.H. Myers, Probability and statistics for engineers and scientists, New York: Macmillan; London: Collier Macmillan, 1989.

[57] D.C. Weaver, C.T. Workman, and G.D. Stormo, "Modeling regulatory networks with weight matrices," Pacific Symposium on Biocomputing, Vol. 3, pp. 112-123, 1999.

[58] P. Yau, R. Kohn, and S. Wood, "Bayesian Variable selection and model averaging in high dimensional multinomial nonparametric regression," Journal of Computational and Graphical Statistics, Vol. 12, pp. 23-54, 2003.

[59] X. Zhou, X. Wang, and E.R. Dougherty, "Construction of genomic networks using mutual-information clustering and reversible-jump Markov-Chain-Monte-Carlo predictor design," Signal Processing, Vol. 83, pp. 745-761, 2003.

[60] X. Zhou, X. Wang, and E.R. Dougherty, "Binarization of microarray data based on a mixture model," Journal of Molecular Cancer Therapeutics, Vol. 2, pp. 679-684, 2003.

[61] X. Zhou, X. Wang, and E.R. Dougherty, "Missing value estimation based on linear and nonlinear regression with Bayesian gene selection," Bioinformatics, Vol. 19, pp. 2302-2307, 2003.

[62] X. Zhou, X. Wang, and E.R. Dougherty, "Gene clustering based on cluster-wide mutual information," Journal of Computational Biology, Vol. 11, pp. 151-165, 2004.

[63] X. Zhou, X. Wang, and E.R. Dougherty, "Gene prediction using multinomial probit regression with Bayesian variable selection," EURASIP Journal of Applied Signal Processing, Vol. 3, pp. 115-124, 2004.

CHAPTER 14

DATA MINING IN BIOMEDICINE

Lucila Ohno-Machado and Staal A. Vinterbo

Decision Support Group, Department of Radiology, Brigham and Women's Hospital, Harvard Medical School

1. Introduction

The study of biomedical processes is heavily based on the identification of generalizable patterns that are present in the data. The practice of medicine is also heavily based on the proper identification of particular patterns of disease progression and the implementation of respective interventions. Data mining algorithms are at the core of the pattern recognition process. In this chapter, we will refer to data mining as any data processing algorithm that has the goal of determining patterns or regularities in the data. These patterns may be used for diagnostic or prognostic purposes and the models that result from pattern recognition algorithms will be refereed to as *predictive models*, regardless of whether they are used to classify (e.g., differentiate among possible diagnoses) or forecast (e.g., estimate a prognosis). This designation will include not only machine learning methods originated primarily from the artificial intelligence and computer science communities, but also regression models originated primarily from the statistical community. Of note, the distinction between statistical and machine learning has become fuzzier in the past few years, as both communities have identified common issues and common solutions to pattern recognition challenges.

We will focus this chapter in the study of data mining motivated by and applied to biomedical data. Biomedical data that serve as inputs and outputs of such models are extremely diverse and multi-modal in their nature, ranging from focused quantitative measurements made at the molecular level to qualitative categorizations used for provision of individual medical care and definition of health care policies. For didactic purposes, we will divide biomedical data mining into two main application areas: (1) biological data mining, including among other applications the analysis of microarrays, construction of phylogenetic trees, analysis of single nucleotide polymorphisms (SNPs) and haplotype tagging, and classification of mass spectroscopy data; and (2) clinical data mining, including the development of individualized diagnostic and prognostic models. We will discuss which types of models are most commonly used in biomedical research, how they are evaluated, and how some are applied in practice. We will review some biomedical applications of predictive models with regards to common issues in their construction, evaluation, and utilization. These three basic steps are also referred to as: (1) derivation, (2) validation, and (3) impact analysis in the clinical literature [McGinn, 2002].

2. Predictive Model Construction

A variety of pattern recognition algorithms have been used in biomedicine with the purposes of exploratory or confirmatory data analysis. The applications span virtually every subspecialty of medicine and the biological sciences. Classification tasks as diverse as diagnosing breast cancer from radiological images [Wei, 2005], analyzing pap smears [Mango, 1998], predicting length of stay in the hospital [Ottenbacher, 2001], determining gene expression profiles [Weber, 2004], making prognoses for patients in the ICU [Beck, 2003], assessing the risk of cardiovascular disease [D'Agostino, 2001], predicting outcomes in spinal cord injury [Rowland, 1998], predicting complications in angioplastic procedures [Resnic, 2001], and diagnosing melanoma from digital pictures [Dreiseitl, 2001].

Unsupervised learning refers to the construction of models that capture the relationships among the data, but that are not guided by the true value of a particular variable (also known as the "gold standard" or "class labels"), which is the case for supervised learning models [Duda, 2001; Hastie, 2001]. We can define the unsupervised learning problem as the problem of *defining* the classes given data without class assignments.

Some examples of unsupervised learning models that have been extensively used in biomedicine are models based on hierarchical clustering, c-means clustering, self-organizing maps, and multidimensional scaling techniques. Examples of the latter are regression models, some types of artificial neural networks, support vector machines, and nearest-neighbors algorithms.

2.1. *Derivation of unsupervised models*

Until recently, unsupervised learning models were seldom utilized in biomedicine. There are currently very few examples of their use in the clinical literature, and those unsupervised learning is to build models that help determine whether regularities in the data can characterize distinct clusters or groups. Such models could have a potential use in the subcategorization of diagnostic or prognostic categories, for example.

In general, we can identify two situations where unsupervised learning is warranted:

1. We are able to obtain samples, but not able to assign one (or more) class labels to each sample, and
2. The characteristics of the class-specific pattern generating system are changing and we need to adapt to these changes.

We can view the class assignment to data points as a function returning a class label when given a data point. The problem is, of course, that this function is unknown to us. Sometimes we are not left without information altogether: we know the general shape or structure of the function involved. What we then need to learn are the parameters of the function involved.

If we are not party to such information, we can use a second approach based on creating a partition of the data into sets or *clusters* according to some overall measure of quality of the partition. Commonly, such measures depend on having defined a *distance* measure d on our data. In general we want d to be such that $d(x,y)$ is "large" when x and y are in different clusters and "small" when they are in the same.

An example of an overall measure is the sum of squared error (SSE) overall measure $J_{SSE}(P)$. If our partition P be into clusters H_1, H_2, ..., H_c, then $J_{SSE}(P) = \sum_{i=1}\sum_{x \in H_i} d(x,m_i)^2$, where m_i is the sample mean (or a representative point as explained below) of cluster H_i. We are now able to outline a variant of the popular c-means algorithm.

1. Choose the number of clusters c
2. *Guess* the initial c means m_i of the clusters
3. Assign each element to the cluster with the closest mean
(iv) Recompute the mean for each cluster resulting from the step 3
(v) If J_{SSE} does not improve, stop and output last clustering, otherwise go to step 3.

One issue with the above approach is that we have to choose the number of clusters c. A way around this requirement is to apply what is known as *hierarchical clustering*. Hierarchical clustering comes in two flavors, bottom up and top down. The bottom up approach initially assigns each element to its own cluster. Subsequent steps merge clusters that are deemed beneficial according to some measure. The top down approach does the opposite and assigns each element to one initial cluster. Subsequent steps then split clusters, again according to some criterion based on a measure. In general these measures are dependent on the distances between elements *inside* of each cluster and the distances *between* clusters, named intra- and inter-cluster distances. Usually we want a combination of minimal intra-cluster distances and maximal inter-cluster distances.

In the following we consider a family of bottom up hierarchical clustering methods. The differences between the members of this family of methods are the different definitions of inter-cluster distances. These are:

1. nearest neighbor, or **single linkage**, where the distance between two clusters A and B is taken to be the minimum distance between two elements, one from cluster A and the other from cluster B.

2. farthest neighbor, or **complete linkage**, where the distance between two clusters A and B is taken to be the maximum distance between two elements, one from cluster A and the other from cluster B.

3. **average linkage,** where the distance between two clusters A and B is taken to be the mean distance between two elements, one from cluster A and the other from cluster B.

Our definition of J_{SSE} can be generalized as m_i can then be assigned according to these three definitions. Once the inter-cluster distance has been chosen, we can apply the following steps for hierarchical clustering.

1. assign each element to its own cluster and compute J_{SSE}
2. merge two clusters with minimal inter-cluster distance
3. stop if only one cluster is left, otherwise go to step 2.

Given the subjectivity involved in assigning meaning to the discovered clusters, and the consequent difficulty in objectively evaluating the results of cluster analyses, unsupervised learning models have not received much attention from the medical research community until the advent of high-throughput microarrays. This was changed because of several articles describing analysis of this new type of data [Eisen, 1998]. An example of hierarchical clustering and a corresponding heatmap is provided in Fig. 1.

Several articles followed the seminal ones and reported on gene expression in different samples, using the same unsupervised learning techniques and software that was made available for this specific purpose. Technical aspects of clustering samples and genes were extensively explored in the microarray literature, but not necessarily compared with supervised learning counterparts. The trend to use unsupervised learning models in microarray analysis is slowly given space to supervised learning models.

Biologists are now used to interpreting dendrograms and expression-based heatmaps generated by hierarchical clustering algorithms and hence they are still utilized for visualization purposes. However, since there often exists gold standard to guide the search for a model, supervised learning models are being increasingly used in this domain.

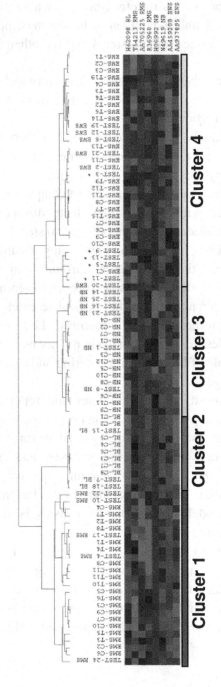

Figure 1: Results of a hierarchical cluster analysis displayed using a dendrogram (tree) and a heatmap. Sample labels are displayed vertically below the leaves of the tree. Variable names are listed on the column at the right side of the figure. Four sample clusters are color coded for easy visualization. The meaning of the clusters is usually assigned subjectively a posteriori, as the main objective of the unsupervised learning algorithm is to indicate the distance (or similarity) among samples based on the values for the variables. In the heatmap, red indicates a low value and green indicates a high value. Gene clusters are listed in the right side of the text column.

2.2. *Derivation of supervised models*

The biomedical literature on applications of supervised learning algorithms is vast, both in biological and clinical applications. The objective of building models using supervised learning algorithms is to predict an outcome or category of interest. For example, it is possible to predict the risk of heart disease for a certain individual within a certain degree of confidence using demographic, history, clinical, and laboratory findings [Ramachandran, 2000]. Models are constructed using data from a large population, such that regularities in the data are captured and applied to new cases. In the microarray data that served as our example in Fig. 1, if clusters of interest are known a priori (e.g., four different types of cancer), then the utilization of a supervised learning algorithm is appropriate.

There are several types of supervised learning algorithms that have been successfully applied in the biomedical literature. It is important to distinguish between algorithms that model continuous outcomes (e.g., prediction of systolic blood pressure, length of survival) versus those that model categorical outcomes (e.g., hypertension, death). Included in the former category are certain types of regression and neural network models. Included in the latter category are logistic or polychotomous regression, certain types of neural networks, nearest neighbors algorithms, classification and regression trees, and support vector machines.

2.2.1. *Statistical regression*

Regression models are very popular in the biomedical literature and have been applied in virtually every subspecialty of medical research. Before computers were widely available, linear regression models were the most popular, as closed solutions to the problem of estimating the intercept and coefficients of the regression equation can be easily calculated. However, models of high complexity have benefited from iterative solutions that are only feasible with the use of computers. For example, logistic regression models, illustrated in Fig. 2(a), model binary outcomes using a logistic function that requires computer-intensive iterative processing to maximize log-likelihood. Several software packages such as SAS [SAS, 2002] implement logistic regression and additionally also have several procedures for model selection (variable selection) and model diagnostics (case selection).

Regression models have the advantage of support from a vast technical and application-based literature, as well as solid foundations in statistical theory. The interpretation of models is more straightforward than that of other types of models, such as neural networks. Linear or semi-linear models, in particular, are less prone to overfitting of memorization of training data, as they are relatively inflexible. This does not mean that they do not overfit the data: in situations in which the number of variables equals or exceeds the number of cases, this is still expected to occur. Popular algorithms for parameter search in this type of models will not run in these cases, justifying the extensive literature in variable selection and principal components regression.

2.2.2. *Artificial neural networks*

An alternative to statistical regression can be found in artificial neural networks (ANNs) [Duda, 2001]. Although originally conceived as a possible model of human cognition, ANNs are widely used to construct predictive models in biomedicine. ANNs may be seen as a superset of regression models, in which the search for parameters is not contrained to the algorithms used for estimating the value of intercepts and coefficients in regression. Furthermore, an essential differentiating factor is that ANNs are more flexible than statistical regression in their ability to model arbitrarily complex functions, since they can incorporate a hidden layer of "nodes" that represent different types of functions between the inputs and the outputs, as well as different types of functions at the output node (and therefore are not constrained to the linear or logistic functions of common regression models), as illustrated in Fig. 2(b). This can be considered both a feature and a drawback in this type of models: more flexibility leads to better fitting, but better fitting may lead to overfitting, the modeling of spurious associations in the data. Therefore, careful control for overfitting needs to be exercised in ANN models.

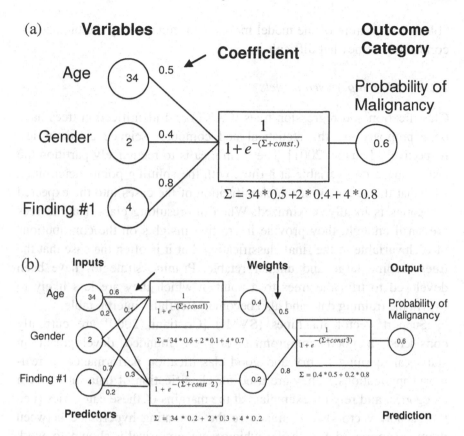

Figure 2: (a) Logistic regression and (b) artificial neural network models for classification of samples into benign and malignant. The classes (or clusters) of interest are defined *a priori*.

The interpretation of neural network models is difficult not only because of the hidden nodes, but also because the incorrect choice of error function may interfere with the semantics of the estimates. For example, if the cross-entropy error function is not used when modeling binary outcomes, the output of the ANN cannot be interpreted as the probability of the outcome coded as "1" (e.g., probability of cancer). Furthermore, since in many situations users not only want to produce a very accurate model to predict outcomes, but also want to gain insight on which factors contribute the most for determining the outcome, the "black box" nature of the model makes its acceptance by the biomedical community somewhat difficult.

2.2.3. Other supervised models

Classification and regression trees (CART) and identification trees have been proposed by the statistical and computer science communities, respectively [Hastie, 2001]. The main idea is to recursively partition the search space one variable at a time, with the splitting points determined such that the accuracy of the classification of the cases into the expected categories is locally maximized. When the resulting classification trees are small enough, they provide interesting insights on the contributions of each variable to the final classification, but it is often the case that the trees become large and uninterpretable. Pruning strategies have been developed to trim the trees to a point in which they are less likely to overfit the training data and also become potentially interpretable.

Support vector machines (SVMs) [Cristiani, 2000] are currently considered a good compromise of well founded development in statistical learning theory and good classification performance in real-world applications. They are used for classification of data into known categories, and rely on exemplars at the margins of these categories (i.e., the support vectors) to characterize the separating hyperplane between them. As opposed to ANNs, which were empirically shown to work before the theoretical foundations were developed, SVMs were the instantiation of learning theory. The basic idea is to transform the input data via kernel functions such that it becomes linearly separable, and identify the support vectors that characterize the best separating hyperplane (i.e., determine the support vectors in which the margin between them and the separating plane is largest). This idea is illustrated in Fig. 3. The kernel functions range from simple polynomials to very complex functions. There is currently little understanding on how to best choose the kernel function for a particular data set, and hence the empiricism of ANNs is also a critical component in the development of an SVM model. Since the first applications in the mid-90s, several examples have been published in the biomedical literature, most notably in biological domains [Furey, 2000].

(a)   **1-Dimensional Data**

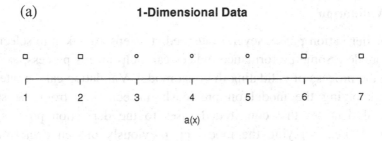

a(x)

(b)   **Data Transformed**

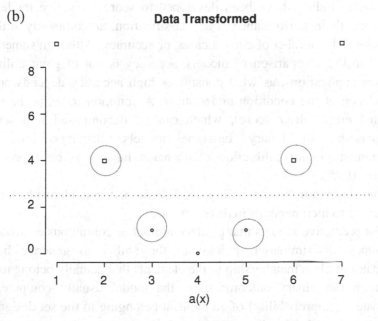

a(x)

Figure 3: One dimensional data for cancer (squares) and non-cancer (circles) cases that have one attribute *a* are shown in Fig. 3(a). These data are not linearly separable. However, if we apply the transformation $T(x) = (x,(x-4)^2)$ to the data, we get Fig. 3(b) where data are separable by the line

$$d = \{(x,2.5) | x \in \nabla\} = \{x | \mathbf{wx} + b = 0\} \text{ where } \mathbf{w}=(0,1) \text{ and } b=-2.5,$$

shown as the dotted line. We can now classify the data item *x* according to whether the point $T(a(x))$ lies above or below the line *d*, or alternatively according to whether $f(x) = sign(\mathbf{w}T(a(x))+b)$ results in -1 or 1. Note that the line *d* is located such that it maximizes the margin between the two classes, i.e., the distance between the points at the class boundaries and the line is maximized. Also note that the location of *d* is dependent only on the points that are circled in Fig. 3(b). These points we call the *support vectors*. In order to avoid computation in higher dimensions, an expression involving a *kernel* function almost always replaces the dot product in the expression for *d*.

3. Validation

In the derivation phase, several intermediate steps are taken to select the best model. Some performance indices can help in this process, as well as in the process of validating the final model. Validation can be internal (i.e., applying the model on previously unseen data from the same original data set that contributed cases to the derivation process) or external (i.e., applying the model on previously unseen data from a different data set.

Several indices have been developed to score predictive models in terms of their performance. For classification, a commonly utilized measure if the number of correct cases, or accuracy. Although sometimes useful in the comparison of models, accuracy is not of great utility in clinical applications, as what constitutes high accuracy depends on the prevalence of the condition under study. A common index is the mean squared error (Brier score), which can be decomposed into several components. In binary outcome models, decompositions into discrimination and calibration components have also been proposed [Arkes, 1995].

Discrimination is the ability to discern between elements labeled according to memberships in different sets.

For predictive models that produce ordinal or continuous estimates of outcomes, discrimination is cast as the ability to generate higher estimates of class membership to the elements that actually belong to that class. In the binary outcome case, the model usually computes its estimate (e.g., probability) of an element belonging to the set designated by outcome "1". The discriminatory ability is then often computed by comparing the relative ranking of estimates for the cases who have an outcome coded as "1" over those for cases coded as "0". This ranking can be measured by the c-index [Harrell, 2001] or the area under the ROC curve [Lasko, 2005]. For example, if "cancer" is coded as "1", then estimates for non cancer patients should be lower than those for individuals who have cancer. Perfect discrimination can be achieved if every non-cancer patient receives an estimate that is lower than that of every cancer patient. However, how close these estimates are to the true underlying probability of cancer for every patient is not assessed in discrimination indices. For this purpose, indices for calibration are informative.

Calibration is the assessment of how close an estimate is to the expected value for the outcome variable. For example, if there are 100 cases that are indistinguishable in terms of their age, gender, and clinical findings, but 90 survive and 10 die, the expected value for the variable "death" (coded as "1") for that type of case is 0.1. A calibrated predictive model would produce an estimate of 0.1 every time it is confronted with a case that matches exactly this profile. Since it is unlikely that cases match perfectly when a large number of continuous variables are utilized, approximations are utilized in indices such as the Hosmer-and-Lemeshow goodness-of-fit statistic for logistic regression models [Hosmer, 2001]. Note that good discrimination is required for these types of calibration measures. Calibration is very important when models are used for individual counseling, as decisions on a particular individual will be based on the probability of outcome determined by the model.

In clinical predictive models, evaluation in terms of areas under the ROC curves is common, and are often accompanied by calibration indices. In biological models, there is still a strong tendency for reporting results in terms of classification accuracy.

A narrow scope in validating a model is its application in a similar setting and population as the ones utilized for model derivation, which is commonly referred to as internal validation. This can be produced with techniques such as cross-validation, bootstrapping, and jackknifing, or prospective cohorts [Efron, 1994]. Neither strategy leads to verification of the usage of the model in other settings or populations. Broad external validation is the evaluation of the predictive model in different clinical settings and different populations. A few clinical predictive models have been subject to this type of evaluation. They are arguably the most widely utilized predictive models in clinical use: the APACHE system for prognosis in critical care [Beck, 2003; Nimgaokar, 2004], and the Framingham model for assessing risk of cardiovascular disease [D'Agostino, 2001]. It is interesting to see these models are both related to medical prognosis, and not medical diagnosis, which have received the most attention from the computer science community in the early 80s. This may suggest that clinicians are less comfortable with the task of prognosis than with the task of diagnosing patients, and therefore may feel that a decision support system for prognostic classification is more useful than a diagnostic one.

4. Impact Analysis

The study of the impact of a model in a particular setting requires several resources. Ideally assessing this impact involves randomized clinical trials in which the predictive model is treated as the intervention. The impact analysis evaluates the overall value of a predictive model in the context in which it will be applied. This is particularly critical in biomedicine, in which medical decision-making may partially depend on the outcome estimates produced by these models.

Evaluating models in terms of the indices described in Section 3 is useful, but provides little insight in terms of added value in the applied setting. Several qualitative and quantitative studies should be conducted to verify how much the information provided by these models adds to the current practice standards. Costs of misclassification (which includes financial costs but also other factors) and benefits need to be taken into account when implementing predictive models in real settings. Answers to the questions below can help guide the impact analysis:

- Is the predictive model going to be used for screening or confirmation of a suspected disease or condition?
- What is the acceptable trade-off between false positives and false negatives in a particular classification problem? Are the estimates from the model dependent on the prevalence of disease in the population?
- Is the model used for develop policies for the population as a whole or to determine follow-up for a particular case at hand?
- Does the model provide information that is currently not available in the environment in which it will be used? For example, are prognoses estimated by computer-based models significantly better than those estimated by health care providers?
- Is the model usable in its current implementation? How does it integrate with the workflow?

5. Summary

Predictive models derived from data mining algorithms have been extensively utilized in biomedical research. In biological applications there are several examples of models derived from unsupervised learning algorithms, particularly in the domain of gene expression microarray analysis. In clinical applications, logistic regression models have been used in practice for prognostication. Several models have been developed in many clinical domains, but few of them have been extensively evaluated in multiple clinical settings using different populations. Two notable exceptions are the APACHE system for prognosis in critical care and the Framingham model for risk assessment of cardiovascular disease. Application in particular settings requires that an impact analysis be conducted, in which the adequacy of the model for the particular situation is assessed in terms of quantifiable costs and benefits.

References

[1] Arkes HR, Dawson NV, Speroff T, Harrell FE Jr, Alzola C, Phillips R, Desbiens N, Oye RK, Knaus W, Connors AF Jr. The covariance decomposition of the probability score and its use in evaluating prognostic estimates. SUPPORT Investigators. Med Decis Making. 1995 Apr-Jun;15(2):120-31.

[2] Beck DH, Smith GB, Pappachan JV, Millar B. External validation of the SAPS II, APACHE II and APACHE III prognostic models in South England: a multicentre study. Intensive Care Med. 2003 Feb;29(2):249-56.

[3] Cristiani N, Shawe-Taylo J. Support Vector Machines and other kernel-based learning methods. Cambridge University Press, 2002.

[4] D'Agostino RB Sr, Grundy S, Sullivan LM, Wilson P; CHD Risk Prediction Group. Validation of the Framingham coronary heart disease prediction scores: results of a multiple ethnic groups investigation. JAMA. 2001 Jul 11;286(2):180-7.

[5] Dreiseitl S, Ohno-Machado L, Kittler H, Vinterbo S, Billhardt H, Binder M. A Comparison of Machine Learning Methods for the Diagnosis of Pigmented Skin Lesions. J Biomed Inform. 2001 Feb;34(1):28-36.

[6] Duda RO, Hart PE, Stork DG. Pattern Classification (2nd ed.), Wiley Interscience, 2001.

[7] Efron B, Tibshirani RJ. An introduction to the bootstrap. Chapman and Hall 1994.

[8] Furey TS, Cristianini N, Duffy N, Bednarski DW, Schummer M, Haussler D. Support vector machine classification and validation of cancer tissue samples using microarray expression data. Bioinformatics. 2000 Oct;16(10):906-14.

[9] Harrell FE. Regression Modeling Strategies. Springer, 2001.
Hastie T, Tibshirani R, Friedman J. The Elements of Statistical Learning. Springer 2001.

[10] Hosmer DW, Lemeshow S. Applied Logistic Regression. Wiley 2001.

[11] Lasko T, Zou K, Bhagwat J, Ohno-Machado L. The use of Receiver Operating Characteristic Curves in Biomedical Informatics. J Biom Inf (in press).

[12] Mango LJ, Valente PT. Neural-network-assisted analysis and microscopic rescreening in presumed negative cervical cytologic smears. A comparison. Acta Cytol. 1998 Jan-Feb;42(1):227-32.

[13] McGinn T, Guyatt G, Wyer P, Naylor CN, Stiell I. Diagnosis: Clinical Prediction Rule. In Guyatt G, Rennie D (eds). Users' Guides to the Medical Literature: A Manual for Evidence-Based Clinical Practice. 2002 AMA press.

[14] Nimgaonkar A, Karnad DR, Sudarshan S, Ohno-Machado L, Kohane I. Prediction of mortality in an Indian intensive care unit: Comparison between APACHE II and artificial neural networks. Intensive Care Medicine, 2004; 30(2):248-53.

[15] Ohno-Machado L, Rowland TR. Neural Network Applications in Physical Medicine and Rehabilitation. American Journal of Physical Medicine and Rehabilitation 78(4):392-8.

[16] Ottenbacher KJ, Smith PM, Illig SB, Linn RT, Fiedler RC, Granger CV. Comparison of logistic regression and neural networks to predict rehospitalization in patients with stroke. J Clin Epidemiol. 2001 Nov;54(11):1159-65.

[17] Ramachandran S, French JM, Vanderpump MPJ, Croft P, and Neary RH. Using the Framingham model to predict heart disease in the United Kingdom: retrospective study. BMJ 2000 320: 676-677.

[18] Resnic FS, Ohno-Machado L, Selwyn A, Simon DI, Popma JJ.Simplified Risk Score Models Accurately Predict the Risk of Major In-Hospital Complications Following Percutaneous Coronary Intervention. Am J Cardiol. 2001 Jul 1;88(1):5-9.

[19] SAS Institute Inc. SAS Online Doc. http://v9doc.sas.com/sasdoc/

[20] Weber G, Vinterbo GA, Ohno-Machado L. Multivariate selection of genetic markers in diagnostic classification. Artificial Intelligence in Medicine, 2004, 31(2), 155-167.

[21] Wei L, Yang Y, Nishikawa RM, Jiang Y. A study on several machine-learning methods for classification of malignant and benign clustered microcalcifications. IEEE Trans Med Imaging. 2005 Mar;24(3):371-80.

CHAPTER 15

MINING MULTILEVEL ASSOCIATION RULES FROM GENE ONTOLOGY AND MICROARRAY DATA

Vincent S. Tseng and Shih-Chiang Yang

Department of Computer Science and Information Engineering,
National Cheng Kung University, Tainan, Taiwan, R.O.C.

Abstract

Data mining methods have been widely used for the analysis of gene expression profiles like microarray data. Among the related studies, association rules mining is a popular technique, by which the hidden relationship among genes can be revealed. In this chapter, we address the issue of combining microarray data and existing biological network to produce multilevel association rules. We use Gene Ontology (GO) to group genes in advance and apply a new multilevel association rule mining algorithm on the hierarchy-information embedded data. The resulted multilevel association rules are assembled by various GO terms that summarize sets of genes, showing the potential relations among gene groups. Through empirical evaluation on real datasets, the proposed approach is shown to be capable of discovering the hidden relations between gene sets in association with GO terms.

1. Introduction

The microarray techniques provide high-throughput capability for revealing valuable information about genes, but the high dimensionality and unstable quality problems in microarray data often bother biologists. In recent years, various data mining techniques have been proposed to solve these problems gracefully. The most popular one among the existing data mining techniques is clustering, which is used to identify groups of genes that share similar expression levels among multiple experiments. The result of clustering can help biologists to discover the genes that may be involved in the same biological process. Association rules, although less frequently used than clustering, have been utilized to find the relationships between genes in some recent work (Berrar *et al.*, 2001; Chen *et al.*, 2001; Creighton *et al.*, 2003; Icev *et al.*, 2003; Kotala *et al.*, 2001). The result of association rule mining is similar to that of clustering, which is group of genes that have some certain relations. However, the meanings of association rules may bring more observations and implications among genes.

An association rule is an expression of the form $X \Rightarrow Y$, where X and Y are disjoint sets of items. The meaning of such a rule is that the Y set is likely to occur whenever the X set occurs. There exist two important metrics for measuring the significance of a rule, called *support* and *confidence*. The *support* of a rule is the frequency that X and Y occur together in a transaction. The *confidence* of a rule is the frequency that Y occurs when X occurs. When applied on the analysis of gene expression data, association rules can reveal some useful implications among genes. For example, a rule G_A (up) $\Rightarrow G_B$ (up) with support x% and confidence c% may show the fact that G_A and G_B are both up-regulated in x% of the experiments. In other words, the rule represents a frequent expression pattern existed in the conducted experiments. The confidence indicates that in c% of the experiments where G_A is up-regulated, G_B is also up-regulated. This measurement displays how closely the expression of G_A and G_B are related. Mining association rules from gene expression profiles, however, tends to generate too many rules for biologists to make use of them.

In a previous study (Tuzhilin *et al.*, 2002), the authors addressed the issue of grouping a huge number of discovered rules produced from microarray data. The ability to group millions of rules enables biologists to analyze the gene groups in which they are interested. However, this grouping method only summarizes the mining result and it cannot reduce the huge number of rules. Hence, there may still be many irrelevant rules. In this chapter, we also use a concept hierarchy to group genes that have the same functions. The difference is that we group genes in advance rather than grouping the resulted rules. The technique used is a variation of association rule, called multilevel association rule.

Multilevel association rule is well introduced (Han *et al.*, 1995), which may lead to the discovery of knowledge in the different concept level of the data. For example, in the marketing analysis, besides finding 80% of customers that purchase jelly may also purchase toast, it could be more informative that 75% of customers buy wheat toast if they buy strawberry jam. The former rule shows the general behavior of customers, but the latter one carries more concrete and specific information. For mining multilevel association rules, a concept hierarchy is provided for generalizing primitive level concepts to high level ones. In this chapter, we use the Gene Ontology (GO) database as the concept hierarchy to group the genes into more general concepts. An example of a multilevel association rule mined from gene expression data with GO might be

$$\{\text{DNA binding}\uparrow\} => \{\text{protein binding}\downarrow\}$$

where the items in the rule are annotated GO terms. This example means that, in most of the experiments where the genes with DNA binding function are highly expressed, the genes with protein binding function are repressed. Our multilevel association rules are similar to the rule groups obtained by the grouping method described in the study (Tuzhilin *et al.*, 2002). However, there exists a significant difference between them. Their grouping method can only group the rules that are found at the gene level. Suppose that a gene group is highly expressed in most of the experiments, but all genes in the group are not highly expressed. Consequently, the rules containing the genes may be treated as insignificant or even not found because of the low support. Multilevel association rules discovered by our approach can avoid this kind of problems.

2. Proposed Methods

2.1. *Preprocessing*

The gene expressions in microarrays are represented as real numbers which are typically normalized according to certain normalization criteria. We discrete these values into three expression levels: *up-regulated* (denoted as ↑), *down-regulated* (denoted as ↓) and *unchanged* (denoted as #). Therefore, the gene expressions in each dataset are formulated as items like gene A↑, gene B↓ or gene C#. In this chapter, we preserve the up-regulated and down-regulated genes, discarding those genes whose expression is unchanged. Since biologists are more interested in those genes that are significantly expressed, which may be more likely to be involved in the regulation of cells than those unchanged genes. Eliminating unchanged genes also reduces the amount of unimportant candidates, thus speeding up the performance. To apply the association rules for mining the microarray datasets, we consider each experiment as a *transaction*. In a transaction, each gene with a discretized expression level is viewed as an *item*.

2.2. *Hierarchy-information encoding*

For the mining of multilevel association rules, Han *et al.* (Han *et al.* 1995) used a hierarchy-information encoded transaction table to replace the original transaction table. This replacement is beneficial because it is inefficient to collect relevant data from files or databases during the mining process. In our work, we also combine the gene expression dataset and the GO hierarchy into a single file so as to eliminate the overhead of querying the GO database in the mining phase.

The Gene Ontology Consortium provides a set of structured vocabularies organized in a rooted directed acyclic graph (DAG), describing the roles of genes and gene products in any organism. Three structured networks of defined terms are provided: molecular function describes the biochemical activity of the gene product, biological process describes the biological goals to which the gene product contributes and cellular component refers to the location in the cell where the gene product exerts its activity. Generally, a gene is annotated by one or multiple GO nodes along the DAG. The nodes at the higher levels correspond to more abstract functional descriptions for gene products. Thus we can view each of the three networks as a concept hierarchy so that we can apply the multilevel association rule mining algorithm.

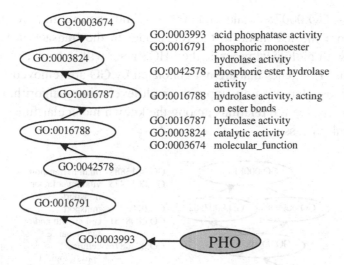

GO:0003993	acid phosphatase activity
GO:0016791	phosphoric monoester hydrolase activity
GO:0042578	phosphoric ester hydrolase activity
GO:0016788	hydrolase activity, acting on ester bonds
GO:0016787	hydrolase activity
GO:0003824	catalytic activity
GO:0003674	molecular_function

Figure 1: The functional path of PHO3 within the molecular function ontology.

With the database of GO downloaded from the GO website (http://www.geneontology.org/), we transform each gene into a path from the leaf node to the root of each branch in GO. Consider the example in Fig. 1. This example shows the functional path of PHO3 in the branch of molecular function. PHO3 is directly annotated by the node GO:0003993, that is, *acid phosphatase activity*. The path from the node GO:0003993 to the root GO:0003674 covers 7 nodes. Thus the item *PHO3↑* in any transaction will be replaced as

(catalytic activity → hydrolase activity → hydrolase activity, acting on ester bonds → phosphoric ester hydrolase activity → phosphoric monoester hydrolase activity → acid phosphatase activity ↑)

when mining within the molecular function ontology.

However, unlike standard trees, the DAG structure of GO allows multiple paths for a given node, especially in the branch of biological process. For example, Figure 2 shows the path of PHO3 within the biological process ontology. The ancestor node GO:0050875 has two parents – GO:0007582 and GO:0009987, thus generating two different paths from the leaf node to the root node. In the transaction table for mining with biological process, the PHO3 gene will be replaced by these two paths. The genes that are not annotated by GO are removed from the transactions. Here we only apply the multilevel mining algorithm on GO terms to observe the relations among the known molecular functions and biological processes.

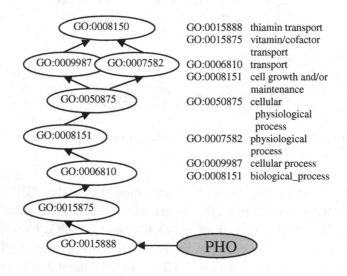

GO:0015888   thiamin transport
GO:0015875   vitamin/cofactor
             transport
GO:0006810   transport
GO:0008151   cell growth and/or
             maintenance
GO:0050875   cellular
             physiological
             process
GO:0007582   physiological
             process
GO:0009987   cellular process
GO:0008151   biological_process

Figure 2: The two paths of PHO3 within the biological process ontology.

3. The MAGO Algorithm

We propose a new data mining method called *MAGO* (*Multilevel Association rules with Gene Ontology*) for mining the gene expression transactions based on the algorithm ML_T1LA proposed in the study of (Han *et al.* 1995). As input, our program accepts the encoded transaction table described above, with items organized by row, and experiments organized by column. The outputs of the algorithm are the rules whose support and confidence values exceed specified minimum support and minimum confidence. The form of rules is

LHS→RHS, support, confidence

where LHS and RHS are 2 disjoint sets of GO terms. The support is the frequency that the rule occurs among the experiments. When LHS occurs in experiments, the confidence is the ratio that RHS also occurs. Therefore, the rules with high support can reveal the general behaviors of the experiments.

3.1. *MAGO algorithm*

Items at higher levels must have higher frequency than their children at lower levels. For instance, the GO term *meiosis* has 9 children – *meiosis I*, *meiosis II*, etc. Genes annotated by *meiosis I* will generate the path that also contains *meiosis* at the encoding step. Thus the frequency count of *meiosis* must be greater than the count of its children. If the minimum support at each level is identical, we may find out the rules that contain items at higher levels, but miss those at lower levels. Obviously, we have to define a mechanism to adjust minimum support at each level, revealing the importance of items at lower levels. In the function *cal_minsup* at line 2, we employ simple criteria to define *minsup[l]* automatically as shown in the following formula:

$$I_{min} = \min(I_1, ..., I_{max\_lv})$$
$$I_{max} = \max(I_1, ... I_{max\_lv})$$
$$\min sup[l] = \max\_s - \frac{I_l - I_{min}}{I_{max} - I_{min}} * (\max\_s - \min\_s)$$

Users must specify two global values – *max_s* (maximum support) and *min_s* (minimum support) for our method to adjust the support value in the given range. The variable I_l represents the number of items at level *l*, I_{max} is the maximum number of items that exist at the same level and I_{min} is the minimum number of items. In the range determined by *max_s* and *min_s*, we adjust the minimum support according to the number of items appeared at the level. As what is shown in the above formula, the level with more items is granted with lower support. Since there are plenty of terms at lower levels of GO, which are more than those at higher levels. Thus low support is reasonable for lower levels to expose the large itemsets among lots of items.

```
         Algorithm MAGO
         Input:  min_lv
                 max_s
                 min_s
                 T
         Output: LL
          1   for(l:=min_lv;L[l,1]≠∅andl<max_lv;l++) do {
          2     minsup[l]:=cal_minsup(l,max_s,min_s);
          3     L[l,1]:=get_large_1_itemsets(T,l);
          4     for(k:=2;L[l,k-1]≠∅;k++) do {
          5       for(cl:=min_lv;cl<=l;cl++) do {
          6         Cₖ:=Cₖ∪get_candidate_set(L[cl,k-1]);
          7       }
          8       foreach transaction t∈T do {
          9         Cₜ:= get_subsets(Cₖ,t);
         10         foreach candidate c∈Cₜ do c.support++;
         11       }
         12       c.support := c.support / |T|;
         13       L[l,k]:={c∈Cₖ |c.support≥ minsup[l]}
         14     }
         15     LL[l]:=∪ₖL[l,k];
         16   }
```

Figure 3: The MAGO algorithm.

Notice that in the generation of candidates, we compute the candidate set C_k from L[0, k-1], L[l, k-1]...L[l, k-1] where l is the current level. This computation relaxes the restriction of mining strong associations among the terms at the same level of GO to allow the exploration of "level-crossing" association rules. Also, level-crossing rules can cover up the indefinite discriminations among levels in GO. For instance, the 2 terms *meiosis I* and *regulation of meiosis* are at the same level of biological process ontology. However, biologists may not care about the relations between them though they are at the same level. Instead, they may be interested in the relation between *meiotic prophase I* (child of *meiosis I*) and *regulation of meiosis*. Hence, we implement level-crossing rules to obtain any relations among GO terms, disregarding the level distinction.

Furthermore, level-crossing rules can expose the distribution of a term in its descendants. For example, suppose we have observed that *meiosis* is up-regulated in most of the experiments, but this GO term is somewhat abstract to us. Then we may want to know among the experiments that *meiosis* is highly expressed, how many of them are also related to *meiosis I*. To answer this question, we simply examine the produced result to find the rule which contains *meiosis* in its LHS and *meiosis I* in its RHS, such as

"meiosis ↑ → meiosis I ↑, support: 10%, confidence: 80%"

The above rule states that *meiosis* and *meiosis I* are up-regulated in 10% of the experiments, and more important, among the experiments that *meiosis* is up-regulated, 80% of them are related to *meiosis I*, the other 20% is related to other descendants of *meiosis* such as *meiosis II* or *regulation of meiosis*, etc. By interpreting this kind of rules, biologists can obtain the further insight into the GO terms in which they are interested.

3.2. *CMAGO (Constrained Multilevel Association rules with Gene Ontology)*

One of the drawbacks of association rule mining on gene expression data is that it produces too many results for biologists to examine, thus losing its usability. Another drawback is that it always filters out those "frequent" rules, therefore, we may find out some frequent but actually well known biological laws, which have no meanings for biologists. Here, we address the issue of filtering the interesting genes in advance, let the mining process focus on the specified GO terms. This kind of rules are named CMAGO (Constrained Multilevel Association rules with Gene Ontology).

We use rule templates to offer biologists the ability to customize their multilevel association rules. A rule template can describe the LHS and RHS terms of the interesting rules. If a rule template is defined, the mining process will only use the chosen GO terms and their children to generate rules. The formalized form of rule template is as follows:

$$(T_1, \ldots\ldots, T_k) => (T_{k+1}, \ldots\ldots, T_n)$$

T_i can be any of the GO terms in GO, describing the GO terms that the resulted rules must contain. At first, the mining process will filter out $T_1 \sim T_n$ and their children terms, to generate large itemsets. In the step of producing rules, the process will select the rules that $T_1 \sim T_k$ occur at LHS and $T_{k+1} \sim T_n$ occur at RHS. For example, assume that we are mining gene expression data with the biological process ontology and we set a rule template as the following:

homeostasis => metabolism

Given this rule template, we will delete the GO terms that do not belong to these two terms, thus limiting the mining scope to the genes described by the two terms. After generating all large itemsets, the mining process will produce rules according to the template. Here is an example:

iron ion homeostasis => amino acid biosynthesis

Because *iron ion homeostasis* is a type of *ion homeostasis* and *amino acid acid biosynthesis* belongs to *metabolism*, this rule conforms to the requirement of the rule template. The other GO terms, even if with very high support, will not be picked out under this rule template definition.

4. Experimental Results

We tested our methods on a cDNA microarray data described in the study of (Hughes *et al.*, 2000). 300 expression profiles were generated in *S. cerevisiae*, each chip containing approximately 6316 genes. We apply the same discretization procedure as the study of Creighton *et al.*, treating an expression greater than 0.2 for the log base 10 of the fold change as up-regulated; a value less than -0.2 as down-regulated, discarding those genes whose expression is between -0.2 and 0.2. We download the three ontology files from GO website, release date is 200405. Then we run the transaction encoding step and produce three files, each integrates the gene expression data and the ontology information. These three files are forwarded to the mining program as input.

4.1. *The characteristic of the dataset*

The three hierarchy-information encoded tables show that the characteristics of the three ontologies are totally different, affecting the subsequent mining process. As shown in Fig. 4(a), some of the 6,316 genes contained in the dataset are annotated by unknown terms. For example, if gene A has not been discovered its molecular function, it will be annotated by *molecular_function unknown* in molecular function ontology. We don't use the unknown terms to generate rules; therefore, the number of rules may be affected by amount of unknown terms.

As we can see, about 1/3 of all genes have no annotations in molecular function, and most of the genes are well annotated in cellular component. Figure 4(b) shows the different GO terms annotated in each ontology. Although most of genes are annotated in cellular component, the number of different terms in cellular component is much less than those in the other ontology. This phenomenon reveals the fact that most genes have the same annotations in cellular component.

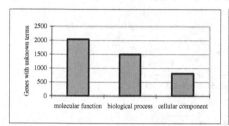

(a) No. of unknown terms

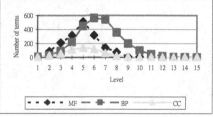

(b) No. of different terms at each level

Figure 4: The characteristic of the dataset.

4.2. *Experimental results*

We setup different *min_s* for the mining process to adjust minimum support dynamically and observe its performance and the amount of rules. The *max_s* value is fixed to 0.7 and the *min_s* is set to 0.1 ~ 0.4. The minimum support at each level is adjusted dynamically in the range determined by *min_s* and *max_s*. Figure 5 shows the mining result with the 3 ontology. As expected, the number of rules and running time increase with the decreasing *min_s*. Even if the annotated genes in cellular component are more than those in biological process, the number of rules in cellular component is much less than that in biological process. This is because the gene annotations in biological process are of more diverse distribution than those in cellular component.

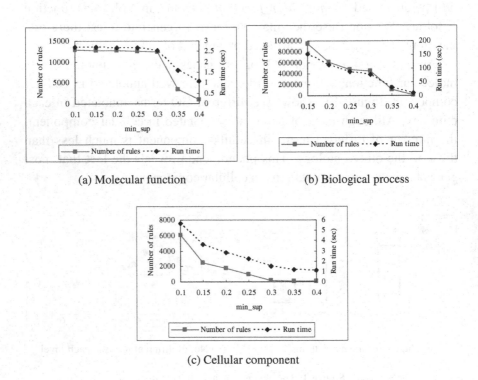

(a) Molecular function   (b) Biological process

(c) Cellular component

Figure 5: The number of rules and run time within the three ontology.

Next we define a rule template and observe its rules under different *min_s* settings. Here we use biological process as the concept hierarchy and define the following template:

metabolism => transport

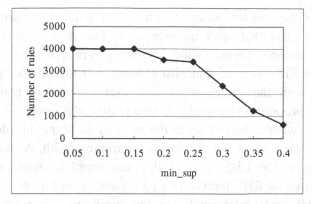

(a) The number of rules generated from biological process with a rule template.

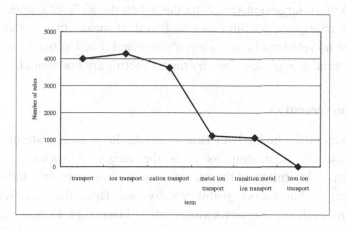

(b) The number of rules generated from biological process with rule templates at different levels.

Figure 6: The number of rules with rule templates.

The above template limits the rules to the genes described by the 2 GO terms. Even if *meiosis* is the most frequent term occurred in the dataset, it will not appear in any of the resulted rules. In addition, the terms under *metabolism* must be placed in the left hand side of rules, and the terms under *transport* must be placed in the right hand side. Figure 6 displays the result of the rule template. Compared with Fig. 5(b), we can realize that the mining process with a rule template generates much fewer rules than that with no templates. The result shows that rule templates certainly focus on the interesting terms, reducing irrelevant rules. From Fig. 6(a), we can also observe that the number of rules is decreased with the increasing *min_s*. Hence the rule template still ensures the properties of association rule mining algorithm.

Next we want to survey when the level of the terms in rule template gets lower, what effect it will bring to the mining result. As well, we use *metabolism* as the LHS term of the rule template, transport and its children terms as RHS terms. As what is shown in Fig. 6(b), when the level of RHS terms becomes lower, the mining process produces fewer rules. This is because lower terms cover fewer children terms, limiting the amount of large itemsets. Note that when the RHS term goes down to *iron ion transport*, no rules can be found. It means that in this dataset there are no relations between *metabolism* and *iron ion transport*, that is, the two gene groups described by the two terms are not related.

4.3. *Interpretation*

As mentioned in the study of Hughes *et al.*, the yeast ergosterol pathway is particularly interesting as it is the target of various antifungal compounds. Inhibition of the pathway causes transcriptional induction of many genes that encode pathway enzymes. Thus, the rules related to ergosterol pathway are worthwhile further inspecting. Here we filter out the rule that contains the term - *"ergosterol biosynthesis"* in its RHS and a single term in its LHS. That is, searching the implications that a group of genes can affect the biosynthesis of ergosterol. There is only one rule that meets our requirement:

tryptophan biosynthesis ↑ => ergosterol biosynthesis ↑ sup: 17%, conf: 75%

This rule indicates that in 17% of the 300 experiments, *tryptophan biosynthesis* and *ergosterol biosynthesis* are both up-regulated. And among the experiments that *tryptophan biosynthesis* is up-regulated, 75% of them exhibit that *ergosterol biosynthesis* is also up-regulated. Such a high support and confidence is extraordinary, we found an article that addressed the relationship between tryptophan and ergosterol (Umebayashi *et al.*, 2003). We, of course, are not suggesting that our rules can uncover the facts exactly, while the precision nature of a rule still requires further investigation in *wet* lab. However, aware of the interpretation of the above rule, biologists can get some directions useful, to identify the mutants or treatments affecting ergosterol biosynthesis.

Individually, the large-1-itemset with the highest support is *meiosis*↑ and *thiamin biosynthesis*↑. These two processes are well known in yeast, and their frequent occurrence is expected. Since yeast is abundant in thiamin (vitamin B1), the high support of *thiamin biosynthesis*↑ certainly illustrates this general behavior. Let us give another example. We searched the Saccharomyces Genome Database website (http://www.yeastgenome.org) and found two saccharomyces pathways – *"homoserine and methionine biosynthesis"* and *"homoserine methionine biosynthesis"*. These two pathways may reveal some relations between homoserine and methionine. In our produced rules, we also find a rule that describes the relationship between them:

homoserine biosynthesis↑ => methionine biosynthesis↓ sup:17%, conf:87%

Inside the rule *methionine biosynthesis* is down-regulated, which brings different considerations to biologists. One may want to know whether there exists any relation of inhibition between the two processes, among the 300 experiments. Besides, there are many rules that contain one of the above two terms, but this rule is granted with the highest confidence relatively. This phenomenon suggests that the association between *homoserine biosynthesis* and *methionine biosynthesis* is strong.

5. Concluding Remarks

Multilevel association rules can reveal the possible genes-to-genes interactions, which also represent the expression patterns occurred in cells. We have proposed a new method for mining multilevel association rules by considering gene expression data and Gene Ontology information simultaneously. The rules discovered by our method can provide biologists insights for investigating the functional relations between gene sets.

Through the proposed mining method, numerous rules may be generated. Most of them are unknown associations that require validation by *wet* lab experiments. However, there exist some relations which are partially validated. For instance, suppose that the relation GOterm A => GOterm B has been verified by biologists. If we derive a new rule such as GOterm A, GOterm C => GOterm B, it indicates a partial set of the validated rule. This kind of rules not only cover the known connections between genes but also reveal new and potential relations between genes. In our future work, we will implement an automatic mechanism to classify the discovered rules into different categories (e.g., unknown, proved and partially proved) so as to expedite the utilization and verification of the mined rules.

Acknowledgement

This research was partially supported by National Science Council, Taiwan, R.O.C., under grant number NSC91-2321-B006-003 and NSC93-2213-E006-030.

References

Agrawal, T. Imielinski, and Swami, A. (1993), "Mining Association Rules Between Sets of Items in Large Databases", In Proc. of the ACM SIGMOD Conference on Management of Data, pp. 207-216.

Agrawal, R. and Srikant, R. (1994). "Fast Algorithms for Mining Association Rules", In Proc. 20th Very Large Databases (VLDB) Conference, pp. 487-499.

Berrar, D., Dubitzky, W., Granzow, M. and Ells, R. (2001). "Analysis of Gene Expression and Drug Activity Data by Knowledge-based Association Mining", In Proc. of Critical Assessment of Microarray Data Analysis Techniques (CAbiDA '01), pp. 23-28.

Chen, R., Jiang, Q., Yuan, H. and Gruenwald, L. (2001). "Mining Association Rules in Analysis of Transcription Factors Essential to Gene Expressions", Atlantic Symposium on Computational Biology, and Genome Information Systems & Technology.

Creighton, C. and Hanash, S. (2003). "Mining Gene Expression Databases for Association Rules", Bioinformatics, Vol. 19, pp. 79-86.

Golub, T. R., Slonim, D. K., Tamayo, P., Huard, C., Gaasenbeek, M., Mesirov, J.P., Coller, H., Loh, M. L., Downing, J., Caligiuri, M. A., Bloomfield, C.D. and Lander, E. S. (1999). "Molecular classification of cancer: class discovery and class prediction by gene expression monitoring", Science, Vol. 286, pp. 531-537.

Han, J. and Fu, Y. (1995). "Discovery of Multiple-Level Association Rules from Large Databases", In Proc. of the 21st VLDB Conference, pp. 420-431.

Hughes, T. R., Marton, M. J., Jones, A. R., Roberts, C.J., Stoughton, R., Armour, C. D., Bennett, H. A., Coffey, E., Dai, H., He, Y. D., Kidd, M. J., King, A. M. *et al.* (2000). "Functional Discovery via a compendium of expression profiles", Cell, Vol. 102, pp. 109-126.

Hvidsten, T. R., Lægreid, A. and Komorowski, J. (2003). "Learning rule-based models of biological process from gene expression time profiles using Gene Ontology", Bioinformatics, Vol. 19, pp. 1116-1123.

Icev, A., Ruiz, C. and Ryder, E. F. (2003). "Distance-Enhanced Association Rules for Gene Expression", 3rd ACM SIGKDD Workshop on Data Mining in Bioinformatics, pp. 34-40.

Kotala, P., Zhou, P., Mudivarthy, S., Perrizo, W. and Deckard, E. (2001). "Gene Expression Profiling of DNA Microarray Data using Peano Count Trees (P-trees)", In Online Proceedings of the First Virtual Conference on Genomics and Bioinformatics.

Pe'er, D., Regev, A., Elidan, G. and Friedman, N. (2001). "Inferring subnetworks from perturbed expression profiles", Bioinformatics, Vol. 17, pp. S215-S224.

Tamayo, P. (1999). "Interpreting patterns of gene expression with self-organizing maps: methods and application to hematopoietic differentiation", Proc. Natl Acad. Sci. USA, Vol. 96, pp. 2907-2912.

The Gene Ontology (GO) Consortium, (2000). "Gene Ontology: tool for the unification of biology", Nat. Genet, Vol. 25, pp. 25-29.

The Gene Ontology (GO) Consortium, (2001). "Creating the Gene Ontology resource: design and implementation", Genome Res., Vol. 11, pp. 1425-1433.

Tuzhilin, A. and Adomavicius, G. (2002). "Handling Very Large Numbers of Association Rules in the Analysis of Microarray Data", In Proc. Eighth Intl. Conf. on Knowledge Discovery and Data Mining, pp. 396-404.

Umebayashi, K. and Nakano, A. (2003). "Ergosterol is required for targeting of tryptophan permease to the yeast plasma membrane", In The Journal of Cell Biology, Vol. 11, pp. 1117-1131

CHAPTER 16

A PROPOSED SENSOR-CONFIGURATION AND SENSITIVITY ANALYSIS OF PARAMETERS WITH APPLICATIONS TO BIOSENSORS

H J Kadim

School of Eng., Liverpool JM University, Liverpool L3 3AF, England, UK

Abstract

Optical biosensors are now utilised in a wide range of applications, from biological-warfare-agent detection to improving clinical diagnosis. An optical biosensor can be sensitive to a number of parameters – external and internal to the sensor. Changes in some parameters may counterbalance changes in other parameters and, hence, the sensor may fail to detect abnormal or unexpected changes in the measurand. Analytical modelling equations to investigate the relationship between the measurand and key parameters – such as temperature, refractive index and time - are presented. These equations can provide useful information for assessing the sensors' performance.

Advances in very large scale integration (VLSI) technology that have allowed the design and manufacture of highly complex integrated circuits at reasonable cost and that form the basis of optical sensors, will compete with the use of stand-alone sensors. A single integrated circuit (IC) is a collection of virtual components, or intellectual property (IP) [1][2]. An IP-core may be a complete communication system, computing system, array of sensors, or a biological laboratory. Hence, sensors have to be able to communicate with neighbouring IPs for better monitoring of an environment or to modify its behaviour, or to interrogate measured information and decide on appropriate action or response. Therefore, to enhance the functionality of sensors, a configuration for a sensor-system is also presented. The realization of the proposed configuration would enhance medical assessment and immediate management of complex conditions.

1. Introduction

The increase in the number of patients, fewer beds, and a possible nursing shortage mean hospitals must increasingly divert critically-ill patients to, for instance, other emergency rooms with the consequent serious threat to patient care. However, it is not always the case that different medical conditions require hospital admission. For example, it has been reported that hospital admission rates for stroke in the United Kingdom is 70% [3]. Stroke accounts for the use of one-fifth of medical beds. This puts a considerable strain on hospitals, as admission for stroke and similar acute health problems is not always necessary. Increasing provision of services in the community can relieve hospitals to treat more serious and urgent cases. This assumes that an appropriate diagnostic assessment has been undertaken with the expectation that the services are part of a specialist health service. Hence, some health related cases could be managed at home with a significant increase in admission and thus hospital beds. This also yields a more effective way of delivery and improving outcome. However, moving acute health problems to home does not come without a cost tag. For instance, the capital and running cost for recruiting, training, and managing caretakers to deal with acute health cases, stretching hospital resources to cover for specialised services outside hospitals an examples. One way of reducing costs but also keeping with the main thrust of releasing more beds to be made available for patients with more complex health problems is by considering a non-invasive sensor-system with wireless link to the hospital computer-network. If such a system is designed whereby the signal(s) of interest, e.g. electrocardiograph (ECG), can be captured, interrogated, and remotely monitored, then patients with heart related problems [4] can be discharged from hospital and treatment can be provided at the point of need. The capability of this sensor-system could be further enhanced such that - after interrogating the captured signal - it can also make a decision on an appropriate action. The appropriate action could be presented to the patient as a set of instructions on a display to be followed [5]. The realization of such sensor-system would enhance medical assessment and immediate management of complex conditions, and improve decision-making at the clinical level.

Furthermore, the establishment of European standards and legislation on water quality has ensured that monitoring has become an essential part of water supply and treatment systems. Monitoring is clearly necessary for ensuring the quality of domestic water supply, but it is also becoming widespread in industrial systems and waste water systems such as sewerage treatment and the outflow of industrial plants. Contamination of water supply may occur in several forms including bacteria and chemical. Accurate measurements of chemical or biological concentration are often essential for monitoring water quality or environmental pollution. There is a clear need for a sensor which is capable of not only monitoring and measuring, for instance, biological agent concentration, but that is able to interrogate measured information and implement control action.

Optical sensors have the potential for making a significant contribution in these areas. The focus of this chapter is two fold: (i) to propose a configuration for optical biosensors which – if implemented - would allow the collection of measured information, interrogation of information and decision making, and (ii) to improve the detection stage of optical sensors by mitigating the effects of internal and external parameter fluctuations, using mathematical modelling equations. This would enable two outcomes: (i) during the design process, the designer will need to consider the effect of parameters on the overall performance of a sensor in detecting unusual events; and (ii) the mathematical equations can be used to optimise the performance of optical sensors by identifying and establishing a relationship between many of the critical parameters that may affect the performance (or response) of sensors.

Optical sensors may differ depending on their area of use, such as for monitoring ligand-receptor binding events, protein adsorption on self-assembled monolayers, glucose, and other types of analyte-surface interactions [6–8]. However, they are all defined by two main elements, an element that creates recognition event and a physical element that transducers the recognition event – e,g. a biosensor can be formed by introducing a recognition layer of biological origin at the optical surface. It is the interaction between this layer and the measurand that will perturb some parameter in the optical-fibre sensor, from which information regarding the measurand can be obtained.

It is not the aim of this chapter to delve into the specific details of the different types of optical sensors (or measurands) and their use, but rather to devise a method of predicting sensor output for a particular measurand condition. The measurand may represent biological or chemical agent concentration, absorbing material, etc.

2. Sensor-System Configuration

The decision for the type of configuration is based on the possibility of future use of a biosensor-system as well as the shape, size and timing of the signal(s) to be processed. Furthermore, it is important to note that sensors no longer perform only a sensing function, but rather they are expected to perform additional tasks, such as interrogation of data, making a decision and, in certain cases, they may be required to control the environment that they monitor. Furthermore, these actions may require to be initiated in real-time with the consequent reduction in time used to implement repair and therefore avoiding possible aggravation of a problem. To meet such requirements, any proposed configuration has to reflect the ever-increasing functional demands for better monitoring of an environment or unusual events. Coupled with future development of system-chip [9] as well as the nature and area of use of a sensor, stand-alone sensors would not meet future requirements.

Figure 1(a) shows a proposed configuration for a biosensor. It comprises a number of software (SW) and Hardware (HW) intellectual property (IP) cores and SW/HW interfaces that enable communication and interaction between embedded cores. Detailed information on the IP cores considered in the proposed configuration is beyond the scope of this book. However, a brief description of their task as well as references will be provided for the interested reader.

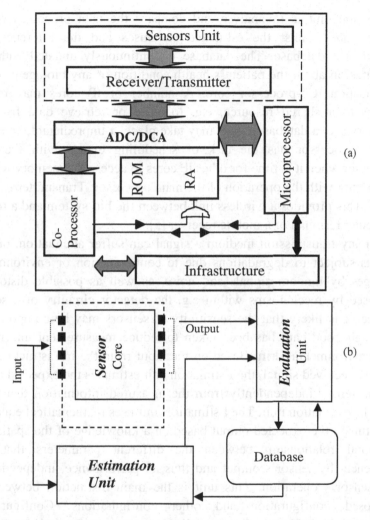

Figure 1: Functional blocks of the proposed configuration; (b) Evaluation/Estimation units.

The main task of the embedded IP-cores is summarised as follows: Sensor unit: This unit encompasses multiple sensors. Each sensor may be designed for a particular functionality, e.g. biological, chemical. Microprocessor: The main task of this core is to implement two distinct SW processes, evaluation and estimation, as shown in Fig. 1(b) The evaluation unit is to analyse and compare the data processed by the sensor-core against the estimated responses and the reference data stored in a database. The database is continuously updated with new information about the patient's health condition or any bio-agent related information. Co-processor core: A number of IP cores may require access to a shared resource, e.g. to write or retrieve data from the memory (i.e. a database). This may take place at unpredictable intervals. The coprocessor has the task of scheduling events with the main processor when it is time for other IP cores to access the memory without interfering with the operation of the main processor. Transmitter/receiver Core: This provides a wireless link between the bio-system and a remote computer terminal or a mobile receiver [5].

In any transmission medium a signal can suffer attenuation, or loss, and is subject to degradations due to contamination or environmental changes by random signals and noise, as well as possible distortions imposed by mechanisms within e.g. the detector circuitry of a sensor. Hence, it is likely that the sensitivity of sensors may be compromised, although good care has been taken to reduce measurement uncertainty and environmental drift. Based on the input from the infrastructure-core and the received signal, the Estimation unit estimates the expected output of the sensor independently from the measured information that is fed into the evaluation unit. The Estimation unit uses mathematical equations to estimate the expected output based on a knowledge of the spatial and temporal relationship between the different parameters that may influence the sensor's output and thus its performance, independent of the sensor's operation. This unit is the main distinction between the proposed configuration and other configurations. Configurations presented in [10], which are realised for a specific task, also have the ability to process and interrogate the measured information with reference to pre-stored data. However, they assume accurate measurements provided by the sensor's detector. This is not always the case as different parameters affecting the measurand may impede each other and thus no noticeable change could be identified by the sensor. The Estimation unit in the proposed configuration, which is populated with information derived from mathematical modelling, provides additional reference data against which the measured information can be assessed and, hence, improve accuracy.

Infrastructure Core: The internal structure of each IP-core involves H/W and S/W implementations, i.e. mixed-signal integrated circuits and software programs. Hard and soft faults [11][12] are likely to occur during the operation of, for instance, a digital-to-analogue (ADC) converter that converts an analogue signal into digital to be processed by the microprocessor. Similarly, software errors or bugs [13][14] may yield wrong results.

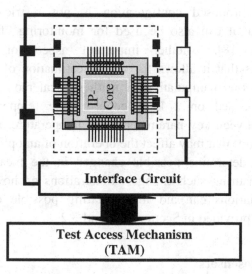

Figure 2: Test access configuration.

The infrastructure-core has the task of implementing regular checks for the purpose of testing and validating the operation of the different IP-cores, as shown in Fig. 2. Deviations in the functional behaviour of the HW/SW functional blocks are fed into the microprocessor – this information is then used by the Estimation unit. For further information on the concept of Infrastructure-Core, the reader may refer to [15]. For diagnosis purposes and also to reduce the cost of implementing these checks, each SW/HW core is specified in terms of input/output constraints [16], i.e. input and output conditions required by the IP-core to process data and produce the output. Once these input/output conditions have been established, violation of conditions is an indication of an error(s) or malfunction within the boundary delimited by the I/O conditions. This can be applied at the interface-level, the IP-level and the process level.

The above IP-cores can be fabricated into a single integrated circuit for a better diagnosis of health related problems. This allows signals to be captured, interrogated and in some cases the problem could be resolved by the patient without the need for further instructions from the hospital. The proposed configuration is not restricted to medical applications but it can also be used for monitoring of environmental pollution [17, 18]. Another important aspect of the proposed configuration is that it allows for virtual estimation of changes in the measurand via the manipulation of mathematical modelling equations. For instance, based on a knowledge of the temporal and spatial relationship between key parameters (e.g. temperature, refractive index of the measurand) that may affect the operation of an optical biosensor, it is possible to determine possible changes in the measurand without physically measuring such changes. Illustrations on how mathematical modelling equations can aid in estimating possible changes in the measurand are provided in Sections 3.1 and 3.2.

3. Optical Biosensors

Biosurveilance is the automated monitoring of information sources of potential value in detecting, for instance, an emerging epidemic. A biosensor usually consists of a bio-recognition element(s) and a transducer. The type of the transducer depends upon the parameters being measured, e.g. electrochemical, optical or thermal changes. An increasing number of biosensors currently rely on the optical properties to monitor, quantify and measure interactions of, for example, bimolecular by coupling sensing elements to a physical transducer. Optical biosensors can be considered as a means of generating light of a particular wavelength, channelling that light into a sample, and then measuring the intensity of light that passes through the sample. Light is introduced into the sample via a waveguide coated with a biological layer. The coated element defines a short range sensing volume within which the evanescent energy may interact with the sample that attenuates it by means of refractive index changes, absorption or scattering [19][20]. Hence, the transmission of light through the waveguide can then be measured as a function of the attenuated field [21] [22], with the profile of the measured light is used to monitor the changes in the behaviour of the attenuated field as a result of changes in the refractive index of the sample [23].

Optical techniques that have come to recent commercial fruition are based on evanescent field sensing [22, 24, 25]. These techniques have a similar optical effect, i.e. the associated electromagnetic field of the light channelled along a surface or reflected within a waveguide penetrates some distance away from the surface into the measurand, e.g. chemical material, before decaying completely. Sensors based on this effect are sensitive to a number of parameters, external (e.g. temperature, detector circuitry) and internal to the sensor (e.g. waveguide properties). The effect of changes in some parameters may counteract the effect of others and, hence, it is likely that the sensitivity of sensors may be compromised, although good care has been taken to reduce measurement uncertainty and environmental drift. Investigation of sensitivity of parameters can aid sensor designers to measure detection system performance. The effect of parameter changes on the sensor's output can be investigated using either direct or indirect method. The former considers the direct effect of parameters on the propagated light and, hence, the sensor's output, whereas the latter considers the behaviour of the electromagnetic field in the measurand [26].

3.1. *Relationship between parameters*

Identifying a relationship between different key parameters affecting the optical-waveguide transmission characteristics can aid in determining the degree of effect of parameter changes on the measured light-intensity. Conversely, knowing the variations in the light intensity it is possible to estimate the degree of effect of these parameters, and thus the changes in the measurand (e.g., variation in biological agent concentration) associated with these changes.

The sensitivity 'S' of the function f(x) to parameter changes can be represented as follows:

$$S_{a_i}^f = \frac{\partial f(x)/f(x)}{\partial a_i / a_i}\bigg|_{\lim_{\Delta a_i} \to 0} \tag{1}$$

a_i: parameter of interest

The following is an example of establishing a relationship between the waveguide's radius, refractive index and the numerical aperture of an optical sensor.

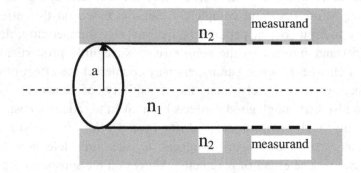

Figure 3: Optical waveguide and the measurand.

The generalised frequency V is the product of a geometric factor $2\pi a/\lambda$ and an optical factor NA, as shown in Eq. (2)

$$V = \frac{2\pi a}{\lambda_o} \sqrt{n_1^2 - n_2^2} \tag{2}$$

where λ_o is the vacuum wavelength, 'a' is the waveguide's radius; n_1 and n_2 are the refractive indices of the waveguide and the measurand. Applying Eq. (1) to Eq. (2):

$$S_V^{n_2} = -\frac{n_2}{V} \times \frac{\pi a}{\lambda_o} \frac{n_2}{\sqrt{n_1^2 - n_2^2}} \tag{3}$$

with

$$\Delta V = -\frac{a\pi}{\lambda_o} \times \frac{n_2^2}{\sqrt{n_1^2 - n_2^2}} \times \frac{\Delta n_2}{n_2} \tag{4}$$

Example: For a 2% change in n2

$$\Delta v = -\frac{\pi a}{\lambda_o} \times 0.07 \bigg|_{n_1=1.5, n_2=1.4}$$

Since the number of modes 'm' propagating into a waveguide is proportional to the normalised frequency, i.e. $m \propto V$, and from (3), growth in the measurand (i.e. an increase in biological or chemical agent concentration) yields a reduction in the number of modes propagating into the waveguide. From (2):

$$\frac{V^2 \lambda^2}{4\pi^2 a^2} = n_1^2 - n_2^2 \tag{5}$$

Re-arranging

$$V^2 = 4(\frac{\pi^2 a^2 n_1^2}{\lambda^2} - \frac{\pi^2 a^2 n_2^2}{\lambda^2}) \tag{6}$$

In the presence of slight changes in the radius of the waveguide and the measurand refractive index:

$$\frac{\partial V}{\partial a \partial n_2} = \pi n_1^2 \partial a - \pi n_2^2 \partial a - 2a\pi n_2 \partial n_2$$

It was shown in [25] that changes in the refractive index of the measurand have an opposite effect on the light-intensity, compared to changes in the waveguide. Hence, at some point, the resultant effect of changes in 'a' and the refractive index of the measurand has a negligible effect on the light-intensity. Therefore, for zero change in the intensity of the measured light:

$$0 = \partial a\pi(n_1^2 - n_2^2) - a\pi n_2 \partial n_2$$

Hence

$$\frac{\partial a}{\partial n_2} = \frac{n_2 a}{n_1^2 - n_2^2}$$

or

$$\frac{\partial V}{\partial n_2} = -\frac{a n_2}{n_1^2 - n_2^2} \frac{\partial V}{\partial a} \tag{7}$$

For zero change in V in the presence of changes in both 'a' and 'n$_2$'

$$\frac{a n_2}{n_1^2 - n_2^2} = 1 \tag{8}$$

with

$$a = (n_1^2 - n_2^2)/n_2 \tag{9}$$

or

$$n_2 a = NA^2 \tag{10}$$

    Equation (8) estimates a relationship between the radius of the waveguide 'a' and the refractive index of the measurand 'n$_2$', that is necessary to maintain constant light-intensity in the presence of parameter changes.

    From (10), knowing the radius and the NA, it is possible to estimate the refractive index of the measurand and visa versa, independent of the sensor's operation. This can also be applied to predict other parameters e.g. spatial parameters of the light source.

For another illustration to estimate the effect of changes in the measurand on the magnitude of the electrical field consider the following:

The decay of the electric field with distance is given by

$$E = E_o e^{-z/d} \quad (z \text{ direction})$$

$$d = \frac{\lambda_o}{2\pi n_1 \sqrt{\sin^2 \theta - \frac{n_2^2}{n_1^2}}}$$

where θ is the angle of incident at the interface of the waveguide.

$$\frac{\partial d}{\partial n_2} = + \frac{\lambda n_2}{2\pi n_1^3} (\sin^2 \theta - \frac{n_2^2}{n_1^2})^{-3/2} \tag{11}$$

$$S_d^{n_2} = + \frac{n_2^2 / n_1^2}{(\sin^2 \theta - (n_2^2 / n_1^2))} \tag{12}$$

as

$$\Delta d \quad \alpha \quad S_d^{n_2}$$

The positive sign in (12) indicates an increase in the magnitude of the electric field in the measurand due to changes in the refractive index of the measurand (e.g. changes in biological agent concentration).

The above equations show examples of a direct method for establishing a relationship between parameter fluctuations, which are influenced by the level of biological agent concentration, and their effect on the light intensity.

3.2. *Modelling of parameters*

It is not always necessary to identify an increase or decrease in chemical or biological agent concentration, or the occurrence of unusual events in the measurand, but rather more importantly the medium's characteristics.

With reference to optical biosensors, the rate and level of changes in the measured light intensity at the sensor's detector – in response to parameter changes – are greatly influenced by the type and concentration of the biological agent. If the parameter changes are investigated with reference to the attenuated field in the measurand, then it is possible to gain an insight into the nature of the measurand, e.g. how it influences the rate of change in parameters. The response of the measurand to certain parameter changes may mask the response to other parameter changes and hence no noticeable related change in the measured light intensity. This indicates that the detection circuitry of the sensor may fail to detect such changes. Populating the Estimation unit in the proposed configuration with such useful information can act as a reference to improve the detection aspect of the sensor.

Fluctuations in one or more parameters may result in a part of the electromagnetic energy is lost with the consequent decrease in the amplitude of the guided wave. This means that some of the light rays are lost and, hence, cannot reach the detector on the other end of the sensor's waveguide. This could be interpreted as that the average transmitted power between the light source and the light-detector is decreased by a factor of $e^{-\alpha z}$ (α: attenuation constant)

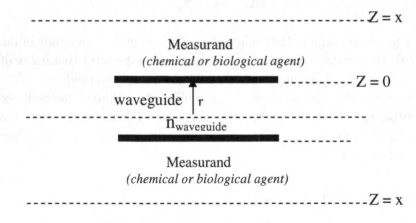

Figure 4: Optical waveguide and the measurand.

With reference to Fig. 4,

$$P_R\{P_{SL}, P_{Act}\} = P_t e^{-b\alpha z} \tag{13}$$

where P_R: power at detector, P_{SL}: power lost in the measurand; P_{Act}: power loss inside the waveguide (i.e. the active region of a biosensor), b: constant

The principle mathematical equations developed for modelling parameters are given in Eqs. (14) to (16), [26].

$$\Re_{P_{Lo}}^{P_{Lz}}(s) = s\,\Re_{P_{Lo}}^{P_{Lz}}(0) + \frac{n_{\tilde{\rho}}}{n_{\tilde{z},\rho}}\,\Re_{P_{Lo}}^{P_{Lz}}(0) \tag{14}$$

where

$$\Re_{P_{Lo}}^{P_{Lz}}(0) = 1\Bigg/\left(s^2 + \frac{\gamma_{j_z}}{\gamma_i}\alpha_{n_z}s + \frac{n_{\tilde{\rho}}}{n_{\tilde{z},\rho}}\frac{\gamma_{j_z}}{\gamma_i}\alpha_{n_z}\right) \tag{15}$$

$$s = -\frac{a}{2}\left(1 \pm \sqrt{1 - \frac{4b}{a}}\right)\Bigg|_{a=\frac{\gamma_{jz}}{\gamma_i}\alpha_{nz};\,b=\frac{n_{\tilde{\rho}}}{n_{\tilde{z},\rho}}} \tag{16}$$

where z: penetration distance in the measurand over which the attenuated field becomes zero. P_{LO}: power at z=0 (i.e. the attenuated field is at its maximum amplitude), P_{Lz}: power at distance z = x (i.e. attenuated field approaching zero; x: integer, x>0). S: complex variable; n: refractive index of the measurand. γ_i and γ_j are parameters representing fluctuations in temperature at z =0 and z = x; $\Re$: transfer function; α_{no}, α_{nz} attenuation factor at z =0 and z= x; a, b: constants. $n_{\tilde{z},\rho}$: fluctuations in refractive index caused by the measurand's parameters (e.g. absorption) with respect to distance and time; $n_{\tilde{\rho}}$: fluctuation in 'n' of the measurand in relation to time only.

With reference to (14), the functional behaviour of the attenuated field in the measurand consists of the original functional behaviour of the attenuated field plus its derivative. If the ratio between the refractive indices is large the functional behaviour is a scaled version of the original behaviour. Otherwise, the derivative term contributes more to the response and has a greater effect.

Equation (14) shows a number of parameters that may affect the attenuated field in the measurand and thus the transmitted light. A sample of the simulation results is shown in Fig. 5.

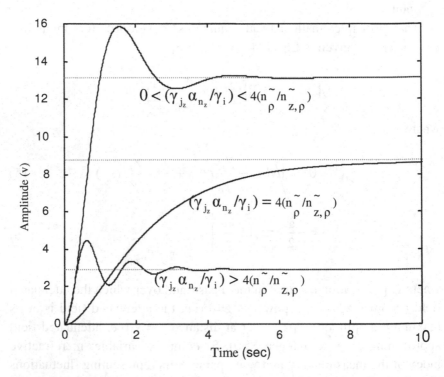

Figure 5: Attenuated field in the presence of parameter fluctuations for a fixed value of z and t (where $t \in \rho$):or a fixed value of z only.

Equation (14) represents parameter fluctuations in the measurand as a function of both distance and time. This means that time (temporal) and distance (spatial) produce different change-signature in biological agent concentration in the measurand.

For similar or comparable spatial and temporal change-signature, (14) can be re-written as in (17)

$$\mathfrak{R}_{P_{Lo}}^{P_{Lz}} = \frac{s+1}{(s^2 + \frac{\gamma_{j_z}}{\gamma_i}\alpha_{n_z}s + \frac{\gamma_{j_z}}{\gamma_i}\alpha_{n_z})}\Bigg|_{\frac{n_{\tilde{\rho}}}{n_{\tilde{z},\rho}}=1}$$ (17)

From (17), the value of s in the numerator is independent of changes in the refractive index. Hence, such changes have no effect on the amplitude of the function, as indicated by the sensitivity results in (18). A sample of the simulation results is shown in Fig. 6.

$$S_{\gamma_i}^{\mathfrak{R}} = +\frac{\gamma_{j_z}\gamma_i^{-1}\alpha_{n_z}}{\gamma_{j_z}\gamma_i^{-1}\alpha_{n_z}} = +1\Bigg|_{s\to 0} \quad S_{\gamma_{j_z}}^{\mathfrak{R}} = -\frac{\gamma_{j_z}\gamma_i^{-1}\alpha_{n_z}}{\gamma_{j_z}\gamma_i^{-1}\alpha_{n_z}} = -1\Bigg|_{s\to 0}$$ (18)

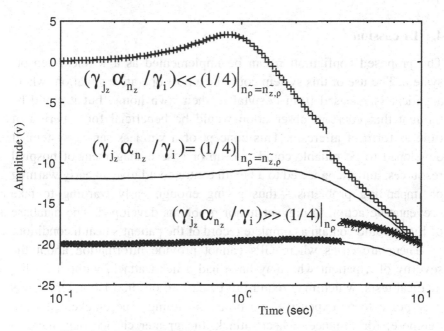

Figure 6: Attenuated field in the presence of parameter fluctuations.

The mathematical equations shown above encompass a number of parameters, which are of spatial and temporal nature. These equations provide us with a means of assessing the behaviour the measurand as well as estimation of the expected response of a sensor to changes in the measurand, independently from the sensor's operation. They provide us with useful information for pre and post design of sensors. In the pre-design case, the mathematical equations can assist sensor designers to predict the expected performance of a sensor. Hence, they can aid in verification of sensors, and also for real-time testing of sensors' performance. Furthermore, for given parameter fluctuations, if a sensor exhibits a noticeable deviation from the expected performance, then the design rules have to be re-visited and updated for better sensor's performance. In the case of post-design, the analytical modelling equations can be used as reference against which the measured information can be checked thereby improving accuracy (see Estimation unit, Fig. 1).

4. Discussion

The proposed configuration can be implemented as a wireless sensor-system. The use of this system could be a post operative situation, where a patient is released from hospital to their own home, but it could be thought that constant observation would be beneficial for a period of time as form of aftercare. This concept of a wireless sensor-system, if developed to be reliable could free up or ease management of hospital resources, and if connected to a health centre could give an early warning of impending problems - thus giving enough early warning to take preventative action to prevent major problems developed. The database of Fig. 1 could contain a complete record of the patient's health condition.

There are cases where GPs cannot provide information about the severity of a patient who may have had a heart attack, without further medical tests, which may require to be carried out in a laboratory – this may prove to be expensive and time consuming. The quicker GPs can diagnose, for instance, a heart attack, the greater chance they have to save a life. Based on the proposed configuration, a patient's medical condition can be monitored and interrogated continuously by incorporating the appropriate sensor in the Sensors unit of Fig. 1 where, for instance, heart-rate related signals can be captured, processed and communicated to a hospital computer-network [5] - thus providing GPs with the ability to detect the presence and strength of a health related problems instantly.

Furthermore, some hospitals are facing serious shortages in their health care workforce, including nurses, radiology and laboratory science and others. The shortage of, for instance, nurses has created not only quality concerns for hospitals, but is also a serious driver of costs, e.g. raise of salaries, recruitment of foreign-nurses. The proposed configuration allows health related problems to be dealt with, with a minimum intervention from hospital and a reduction in admission and lengths of stay for relatively healthy patients – thus releasing more hospital resources. It can also help in post-care monitoring and evaluation, e.g. identifying patients whose progress has deviated from the norm after they are sent home. Based on the proposed configuration, a system-chip can be fabricated and used not only to improving medical diagnosis or detection of abnormal health conditions, but it can also be used in biological warfare detection by simply incorporating the appropriate sensor as a part of the sensor-core in Fig. 1.

Establishing a relationship between changes in the measurand and parameters affecting the performance of a biosensor can aid in verification of sensors. A noticeable drift between the estimated light intensity (Estimation unit, Fig. 1) and that measured at the detector is an indication of abnormality in the sensor's functionality. To ensure consistence performance of sensors a test can be applied at regular intervals. From the mathematical equations it is possible to identify possible changes in the measurand and their corresponding changes in light intensity. If this information is stored in a database, then it can be used to detect changes outside the norm for the measurand, and subsequent feedback and compensation to maintain normal operation.

Furthermore, there will be constraints on what can be achieved, not only limits to the physical capabilities of the optical equipment being used for testing or assessing sensors' performance, but also, for example, economic, legal and safety constraints. The mathematical equations can be realised by electrical circuits which can be used as a model to analyse the behaviour of the attenuated field without the need for the optical arrangement, and hence easy to integrate software and computer simulation. Other advantages can be summarised as follows:

(i)   They can be used to estimate the possible changes in the measurand independent of the sensor's operation.

(ii)   They can provide necessary information that can be used for verification of a sensor as well as for testing purposes.

(iii)   Sensors can form an essential part of a measurement or information processing system. The output from a sensor can be used to drive a process and, in effect, control it. A software algorithm of the mathematical equations can be developed and can easily be integrated into sensors.

(iv)   Mathematical equations may be used to test the effect of various parameters to a real system. This is desirable for many reasons, but mainly because tests on the real system may be costly or impractical.

5. Conclusion

Support and management of patients, with acute health related problems, can result in appropriate treatment and achieve efficiencies by way of reducing unnecessary admission. A configuration for a sensor-system that enables acute health related problems to be dealt with, with a minimum of hospital intervention has been presented. A biosensor system - based on the proposed configuration - can help reduce the rate of admission to hospitals and, hence, releasing more hospital resources. The realisation of the proposed configuration also enables electronic disease reporting of symptoms and diagnostic results that could be shared by different health authorities.

In any transmission medium a signal may suffer attenuation or loss, and is subject to degradations due to contamination by random signals and noise. In addition, the responsitivity of the detector circuitry may vary over the range of operation of the sensor. Therefore, the measured light is less likely to be a true reflection of changes in the measurand. As the modelling equations establish a relationship between the measurand and key parameters affecting the light intensity, they can provide useful information that will help to identify inconsistencies in sensor performance.

References

[1] M.I. Hunt, J.A. Rowson, "Blocking in a System on a Chip", IEEE Spectrum, Nov. 1996, pp.3541.

[2] E.A. Senior, "Test and measurement", IEEE Spectrum, January 2000, pp.75-79.

[3] A. Bhalla, R. Dundas, A. Rudd, C. Wolfe, "Does admission to hospital improve the outcome for stroke patients", British Geriatrics Society, Age and Ageing, 2001, pp. 197-203.

[4] R.M. Berne, M. Levy, "Cardiovascular physiology", Mosby, Inc., USA, 2001.

[5] H.J. Kadim. "Mixed-signal IP-Core for Virtual Monitoring of Health related Problems," 11th International Conference Mixed Design of Integrated Circuits and Systems, Szczecin, June 2004, pp. 585-588.

[6] http://pubs.acs.org/cen/topstory/8033not4.html

[7] http://www.devicelink.com/mddi/archive/97/11/011.html

[8] http://www.fraserclan.com/biosens6.htm

[9] C. Rowen, Engineering the Complex SoC, Pearson Education, Inc., NJ, 2004.

[10] J.W. Gardner, Principles and Applications, John Wiley & Sons Ltd, England, 1994,

[11] H.J. Kadim, "Estimation of a Maximum Bound of Uncertain Parameter Fluctuations with Applications to Analogue IP-Cores", Proc. of International Symposium on System-on-a-Chip, Finland, November 2004, pp161-164.

[12] M. L. Bushhnell, V.D. Agrawal, "Essentials of Electronics Testing for Digital Memory & Mixed-signal VLSI Circuits", Kluwer Academic Pub., USA, 2000.

[13] R.S. Pressman, "Software engineering", McGraw-Hill, Ltd., UK, 2000, PP.433.

[14] J.G. Brookshear, "Computer science, an overview", Pearson Education, Inc USA.

[15] A. Benson, S. Carlo, P. Prinetto, "A Hierarchical Infrastructure for SoC Test Management", IEEE Design and Test of Computers, August 2003, pp. 32-39.

[16] H.J. Kadim. "On the Assertion-based Verification of SoC," 11th International Conference Mixed Design of Integrated Circuits and Systems, Szczecin, June 2004, pp. 230-233.

[17] http://physicsweb.org/article/world/12/2/6.

[18] Y.M.Wong, P.J. Scully, H.J.Kadim, V.Alexiou, R.J. Bartlett, "Automation and dynamic characterisation of light-intensity with applications to tapered plastic optical fibre", Photon 2, Pure and Applied Optics, 2003, pp. s51-s58.

[19] K.T.V.Grattan, B.T.Meggitt, "Optical fibre sensing technology, chemical and environmental sensing (Optoelectronics, imaging and sensing)." (Kluwer Academic Publishers 1999).

[20] C.R.Lavers, Z.Q. K.Itoh, M.Murabayashi, "A composite optical waveguide sensor utilising induced surface scattering, and applications for liquid and vapour sensing". Proc. of Sensors and their application XI. Institute of Physics Publishing, UK, Sept. 2001, 281-287.

[21] M.J. Usher, D.A. Keating, "Sensors and transducers", MacMillan Press, Ltd., UK, 1996.

[22] R. Philip-Chandy, P.J. Scully, H.J. Kadim, M. Gerard Grapin, "An Optical Fibre Sensor for Biofilm Measurement using Intensity Modulation and Image Analysis", IEEE Journal on selected topics in quantum Electronics, Vol.6, No. 5, 2000, pp. 764-772.

[23] G. Stewart, J. Norris, D. Clark, B. Culshaw, "Evanescent-wave chemical sensors", Inter. Journal of Optoelectronics, Vol 6, No. 3, 1991, pp. 227-238.

[24] D. Merchant, P.J. Scully, R. Edwards, J. Grabowski, "Optical fibre fluorescence and toxicity sensor", Elsevier Science, sensors and actuators B48, 1998, PP. 476-484.

[25] Y.M.Wong, P.J. Scully, R.. Bartlett, V.Alexiou, H.J.Kadim, "Automation and Characterisation of Chemical Tapering of Plastic Optical Fibres", Proceedings of the Eleventh Conference on Sensors and their Applications, London, September 2001, pp. 203-208.

[26] H.J. Kadim, "Power Loss under Parameter Fluctuations with Applications to Linking-with-Light System-on-a-Chip", School of Engineering, Liverpool JM University, 2004.

EPILOGUE

Chung-Sheng Li and Stephen TC Wong

Data driven analysis, modeling, and inferencing is perhaps one of the most fundamental methodologies for conducting informatics research in the life sciences. Traditionally, there is a long lifecycle in designing the protocols and experiments, collecting the experimental data, performing analysis on the acquired data and inferencing potential implications. The recent rapid advance of biotechnology, for example, cell level analysis through high throughput techniques, allows collecting large amounts of digital data within a very short period of time. On the other hand, from the clinical aspects of life science, many hospitals have also digitized most of their medical charts and laboratory data and are in the process of standardizing and integrating their patient record systems; this in turn allows, for the first time, large amounts of longitudinal patient data to be within reach for near real-time epidemiology study. These new advances in healthcare technology have definitely accelerated the need for developing new analysis and modeling techniques in the clinical side of life sciences.

There are three reasons why the topics presented in this book edition could be of great interests to the readers who are interested in developing data mining and analytic algorithms and systems for the rapid evolving life science applications:

(1) Bio-surveillance for detecting bio-terrorism (anthrax, small pox), health activity monitoring for detecting disease outbreak, such as Avian flu, West Nile, and SARS, environmental activity monitoring, such as air pollution, and forest fire, have become increasingly important and problematic. The activity monitoring efforts include sensor network-based and syndromic-based approaches. Some of these efforts have received a lot of attention today and have reached a certain level of maturity during the recent years due to substantial investment from government sponsored programs, e.g., DARPA Bio-Alirt and CDC Bio-Sense programs.

361

(2) Large amounts of multidimensional signal and image data have been generated and collected as a result of the rapid adoption of high-throughput techniques in life science experimentations, based on new advances in a broad spectrum of microscopy modalities, biosensors, and microarrays. The data are collected at either the subcellular, cellular, or tissue levels and are used to profile or screen compounds for drug discovery or to study molecular or cellular mechanisms in systems biology fashion. It has been felt by this community that applying quantitative and mathematical techniques of data mining to extract models from the collected and processed data will be able to help us to gain both basic understanding of the causative mechanisms of diseases as well as to translate the results effectively into clinical space for cost-effective diagnosis and disease prevention.

(3) The integration of macro-scale bio-surveillance and environmental data with micro-scale biological data will have the potential to provide new, more robust public and personalized health applications and services than the current approaches that are often based on a single scale of biological evidence.

This book intends to identify and highlight new data mining paradigms to analyze, combine, integrate, model, and simulate vast amounts of heterogeneous multi-modal data for emerging life science applications. In contrast to alphanumeric data generated in genetics and genomic studies, these multi-model and multi-dimensional data (Chapters 2, 4, 11-13, 16) are generated by the availability of high throughput, large scale imaging or signaling acquisition devices, including various microscope modalities (e.g., fluoresence, light, optical, confocal, 2-photon laser scanning, automated, time-lapse, laser capture dissection, and electronic), biosensors, and microarrays. This renders the biomedical community, for the first time, facing the prospect of identifying and understanding the functions and interactions of macromolecules in cells and decoding the biological mechanisms of diseases and organ development, as well as the translation of the basic scientific discovery effectively into clinical space, including diagnosis, prognosis, therapeutic prevention, repair, drug development, and disease management [1].

At a more macro level, monitoring disease outbreaks for bio-surveillance and public health purposes based on environmental epidemiology has been demonstrated for a number of vector-born diseases, such as Hantavirus Pulmonary Syndrome (HPS), malaria, and Denge fever. Recently, health activity monitoring (HAM) concept has also been applied to the early detection of subtle human behavior changes due to disease outbreak to provide advanced warnings before significant casualties registered from clinical sources (Chapters 6-9). The alerts generated from HAM systems can be triggered through the fusion of both traditional and non-traditional multi-modal heterogeneous data sources. Traditional data includes data generated from clinical sources, such as inpatient and outpatient data. Non-traditional data sources include those data collected from remote sensing (including satellite images), video/audio surveillance, home telemetry, and other data to enable the possibility of extrapolating or predicting population behaviors.

More specifically, this book intends to address various aspects of the emerging research topics in life science data mining. As we have most evidentally shown in this book, these aspects could include:

- **Granularity**: from genes (Chapter 15), to cells (Chapter 2, 12), to individuals (Chapter 5), and to public and population health (Chapter 6-9);
- **Data modality**: patient data (Chapter 6-9), EEG (Chapter 5), high throughout acquisition devices such as microarray and automated microscopy (Chapter 2, 4, 11-13), and optical biosensor (Chapter 16)
- **Problem domains**: flu (Chapter 7), substance abuse (Chapter 9), cellular and signaling pathways of various human diseases (Chapter 2, 4, 11-13);
- **Data mining techniques**: diversity (Chapter 3), clustering techniques (Chapter 4), parametric modeling (Chapter 5), Bayesian Fusion (Chapter 6), ensemble classification (Chapter 11), predictive models (Chapter 14), and data mining association rules (Chapter 15).
- **Knowledge representation**: Predictive Toxicology (Chapter 10).

In summary, this book by no means has exhaustively investigated the relevant aspects of life science data mining. Nevertheless, we believe that this collection of reported work from the leading researchers in the field today would offer the readers a reasonable good start in these areas. Furthermore, we also hope to use the book as a vehicle to promote and motivate possible sharing of similar data mining techniques across multiple granularities, data modalities, and problem domains in life sciences.

References

[1] J. Chen, E. Dougherty, S. Demir, C. Friedman, CS Li, and STC Wong, Grand Challenges for Multimodal Bio-Medical Systems, *IEEE Circuits and Systems Magazine*, vol. 2, No. 2, pp. 46-52, May 2005.

INDEX